1.75

NURSES' AIDS SERIES

Anatomy & Physiology for Nurses

NURSES' AIDS SERIES

ANAESTHETICS FOR NURSES

ANATOMY AND PHYSIOLOGY FOR NURSES

ARITHMETIC IN NURSING

EAR, NOSE AND THROAT NURSING

MEDICAL NURSING

MICROBIOLOGY FOR NURSES

OBSTETRIC AND GYNAECOLOGICAL NURSING

ORTHOPAEDICS FOR NURSES

PAEDIATRIC NURSING

PERSONAL AND COMMUNITY HEALTH

PHARMACOLOGY FOR NURSES

PRACTICAL NURSING

PRACTICAL PROCEDURES

PSYCHIATRIC NURSING

PSYCHOLOGY FOR NURSES

SURGICAL NURSING

THEATRE TECHNIQUE

TROPICAL HYGIENE AND NURSING

NURSES' AIDS SERIES

■ EIGHTH EDITION ■

ANATOMY & PHYSIOLOGY FOR NURSES

KATHARINE F. ARMSTRONG ✠

SRN, SCM, DN(Lond)

Revised by

SHEILA M. JACKSON

SRN, SCM, BTA, RNT

*Inspector of Training Schools
for the General Nursing Council
for England and Wales.
Formerly Nurse Tutor,
Queen Elizabeth School of Nursing, Birmingham
and St Charles' Hospital, London*

With a Foreword by

E. W. WALLS

MD, BSc, FRSE

*Courtauld Professor of Anatomy,
The Middlesex Hospital Medical School,
University of London, London*

BAILLIÈRE ⚜ TINDALL · LONDON

© 1972 BAILLIÈRE TINDALL
7 & 8 Henrietta Street, London WC2E 8QE

A division of Crowell Collier and Macmillan Publishers Ltd

First edition, 1939
Second edition, 1941

Seventh edition, 1964
Reprinted, 1969, 1972
Eighth edition, 1972
Reprinted, 1974, 1975, 1976, 1977

ISBN 0 7020 0436 7 Limp edition

E.L.B.S. edition, 1973 (ISBN 0 7020 0493 6)

Sinhala edition (Official Language Affairs, Colombo), 1958
Turkish edition (Turkish Government), 1960
Spanish edition (C.E.C.S.A., Mexico), 1966
Reprinted, 1969, 1972

Published in the United States of America by
the Williams & Wilkins Company, Baltimore

Made and printed Offset Litho in Great Britain by
Cox & Wyman Ltd,
London, Fakenham and Reading

Foreword to the Eighth Edition

At the present time essentially the same problem faces those who have to plan the training of nurses as confronts those of us who are primarily concerned with producing the nation's doctors; in a word, how best to organize the curriculum in harmony with rapid advances in medicine, and with the increasing age structure of the population.

Now, when the end-product is already first-class—and the British-trained nurse needs no accolade from me or anyone else—those who would change things must proceed with caution, for there could be no greater lack of wisdom than to weaken the basic instruction upon which so much depends. No one will dispute that however one sees the nurses of the future their traditional role at the bedside will predominate; but to that role there must be coupled a proper understanding of the meaning and purpose of modern technological procedures, whether diagnostic or therapeutic.

I believe it to be self-evident that such understanding rests upon an adequate knowledge of the anatomy and physiology of the human body, and to say so is not to deny that the word adequate might be the subject of lengthy and heated debate. Nevertheless, I submit that in this well-known and deservedly successful book the correct balance has been struck, and that it will continue splendidly to serve the needs of the nursing profession.

In her preface Miss Jackson has indicated the changes which she felt were desirable to ensure that this new edition would be in line with current knowledge and contemporary teaching methods. She has chosen well, and both the text and accompanying illustrations are a model of clarity; while, of course, to an anatomist it is a special pleasure to welcome the use of the most recent, internationally agreed, anatomical nomenclature.

It is perfectly evident that Miss Katharine Armstrong's book was entrusted to safe hands, and not another word need be said.

ELDRED W. WALLS

The Middlesex Hospital Medical School
London W.1
June 1972

Preface

It was with some trepidation that I undertook the revision of *Anatomy and Physiology for Nurses*, for Miss Armstrong's book is so well known and has proved so popular with nursing learners at all levels that I thought it would be difficult to improve upon the basic structure of her text. However, changes have been necessary to meet the requirements of today's reader and the new approach to the teaching of anatomy and physiology resulting from new knowledge and from consequent changes in the nurses' syllabus. In bringing this book up to date my personal gain has been the amount that I myself have learned, and a renewed realization of the miracle of the human body. I hope that this stimulus and interest will be passed on to other students of this book.

In some instances there has been need for additional material, and some sections have been enlarged in order to clarify a point so that the understanding of recent medical advances should be a little easier. The first three chapters introduce briefly some scientific principles which will help in the understanding of the chemistry of the body in later chapters. In this section, and throughout the book, metric measurements have been used. More sub-headings have been introduced in order that looking up a reference may be as easy as possible. The third chapter contains some additional material on genetics and sex determination; sufficient, it is hoped, to stimulate further study of this fascinating subject. Chapter five has been rewritten, as has the section on blood groups, and paragraphs have been added on the Rhesus factor and on immunity, the mechanism of which must be understood if some modern concepts of disease are to be appreciated. The chapter on the respiratory system has also been re-written and enlarged but at the same time simplified a little.

There has been some rearrangement of the order of the chapters to make study more logical for the student and there is greater emphasis than hitherto on the physiology, or function, of the body. The structure of the book has been influenced by new teaching methods, in particular by the techniques of programmed learning, which enable the student to progress through the work in logical sequence, one step at a time. Although it is of course impossible to study any one system of the body in isolation, an attempt has been made to build up the knowledge, so that information needed to understand one section has been given in a previous chapter.

I have not marked certain paragraphs for study by advanced students, as was done in previous editions, as the material contained in these sections is nowadays necessary for an understanding of the subject and is not beyond the capacity of the more junior student. Likewise the sections for practical work at the end of each chapter have been omitted because students who have a knowledge of scientific principles will have carried out these experiments, and those who have not studied science require more comprehensive instruction than can be included in a book such as this.

An important change has been the alteration of terminology to the recommended English equivalent of the Paris *Nomina Anatomica*; these terms, though relatively new to some of us, are increasingly used, and I have followed the thirty-fourth edition of *Gray's Anatomy* in this respect. A short glossary is given at the end of text for ready reference to both old and new terminology. I have also included a short bibliography of books which will be of interest to students wishing to gain a little deeper knowledge of certain subjects.

As many of the illustrations required alteration and a number of new ones were to be added, the opportunity has been taken to redraw all illustrations and to make use of a second colour printing. The text has been redesigned to make for easier reading and study.

I would like to thank my colleagues throughout the nursing profession, and in particular tutors at St Charles' Hospital, London and at Queen Elizabeth School of Nursing, Birmingham, for their help and advice—both sought and unsought!—and my thanks are also due to the publishers and their staff for their unremitting help and guidance in the preparation of this edition.

March 1972   S.M.J.

Contents

1 Elementary Physics and Chemistry

THE NURSE cares for human beings in illness and during vulnerable periods of their lives. In order to do this properly she needs to understand the structure and function of the body and to be able to relate this knowledge to her care of the patient. Each organ in the body plays its part in maintaining the health of the whole, and if one organ is at fault the whole body will be affected. The structure of each part suggests the function, and the function suggests the structure, so that the study of the human body is a logical process of thinking and reasoning, not merely of memorizing.

Terms

Anatomy is the study of the structure of the body and **physiology** is the study of its function. Some knowledge of elementary physics and chemistry will help in the study of anatomy and physiology and will also be needed to enable the nurse to care for the patient and to carry out nursing procedures. **Chemistry** deals with the composition of matter and the reactions between various types of matter. **Physics** deals with the behaviour and characteristics of matter, for example whether it gives off heat and light, or conducts electricity.

Matter

Matter is anything that can occupy space. When we refer to a 'space' we usually mean that it contains air, which can be displaced by other forms of matter. For example, an 'empty' bottle is not really empty as it contains air. A container from which the air has been extracted is called a vacuum. Matter has three physical states: (a) solid, (b) liquid, and (c) gas.

A solid does not alter easily in either shape or size, e.g. stone, brick.

A liquid takes the shape of the vessel it is put into, but does not alter in size, e.g. 500 millilitres (ml) of water has no shape, but it takes the shape of the container it is put into; if it is put into a jug which will hold 1 litre the jug will not be filled.

A gas takes the shape and size of the vessel containing it. If the air is extracted from a container and a small quantity of gas is introduced the gas will expand to fill the container. If more gas is introduced the gas in the container is compressed, and the small particles which make up the gas are packed more closely together. More gas could be introduced until the flask bursts. A common example of this can be seen when a bicycle tyre is pumped up. After a few pumps the tyre seems full, but when someone gets on the bicycle the tyre goes flat because there is so little air in it. You can go on pumping until there is so much air in the tyre that when you get on the bicycle your weight does not flatten the tyre. A measured amount of gas in a cylinder will exert a known pressure. As the gas is used up the pressure in the cylinder will drop and it is in this way that the quantity of gas remaining in the cylinder can be gauged. Common examples of gases are oxygen, hydrogen and nitrogen.

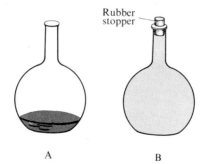

FIG. 1. Glass flasks each containing 100 millilitres of (A) fluid, and (B) gas. Note how the gas expands to fill the flask.

Change of State

At normal temperatures iron is solid and water is liquid, but the state of matter can alter. The changes are due to heating or loss of heat. For example, liquid water becomes solid ice when heat is lost, and solid butter becomes liquid oil when it is warmed. The changes can be described as follows.

Melting is the change of a solid into a liquid as a result of heating, e.g. the turning of ice into water.

Evaporation is the change of a liquid into a gas as a result of heating, e.g. the turning of water into steam and water vapour. Gases formed by evaporation are called vapours.

Condensation is the turning of a gas into a liquid as a result of cooling, e.g. the turning of water vapour into water.

Consolidation is the turning of a liquid into a solid as a result of cooling, e.g. the turning of water into ice.

The first two of these changes, melting and evaporation, are due to heating, but the heat which causes this change of state does not cause any rise in the temperature of the matter. Heat is a form of energy and the energy in this case is used up in producing the change of state, so it does not heat the matter. For example, the temperature of melting ice is 0°C (32°F). When the ice melts the water is also at a temperature of 0°C. Water boils at 100°C (212°F). If you leave the kettle on the heat the water will not get any hotter because the added heat is used up in turning the water into water vapour. In the same way heat from the body evaporates the sweat from the skin, excess body heat is used up and the body is cooled. Heat used up in this way is called *latent heat*. In condensation or consolidation latent heat is set free.

Analysis and Synthesis

Earlier it was explained that chemistry is the study of the composition of matter. A chemist tries to split matter up into its various components to see what it is made of. This process is called *analysis*. He may also try to build up matter from its various components. This process is called *synthesis*. Synthetic substances are those which are made by man, e.g. plastics and nylon.

Elements, Compounds and Mixtures

Some substances cannot be split up and these substances are called *elements*. There are over ninety elements known to exist, including oxygen, carbon, nitrogen, iron, silver and gold. Substances which can be split into different elements are of two types: (a) compounds, and (b) mixtures.

A compound is a substance made of two or more elements which combine chemically to form a new substance with new properties. For example, hydrogen is a very light gas, used to fill balloons. It is highly inflammable and may explode. Oxygen is a gas which supports combustion but will not itself burn.

These two gases in combination make a compound, water, which is a fluid, but it is not inflammable, will not explode, will not support combustion and will not burn. In fact it is very useful for putting out a fire. The elements which form a compound are always present in fixed proportions. For example, water consists of two parts of hydrogen combined with one part of oxygen. If the proportions were altered a compound might be formed, but it would not be water. Two parts of hydrogen combined with two parts of oxygen give a compound called hydrogen peroxide which looks like water but has very different properties. Another characteristic of a compound is that it is not easy to separate the elements which compose it. To accomplish this separation a chemical change must take place.

A mixture is a substance made of two or more elements mixed together but not chemically combined. Air is a mixture of gases, mainly nitrogen and oxygen, with traces of carbon dioxide and other gases. A mixture has no new properties but simply those present in the elements which form the mixture. Air supports combustion because it contains free oxygen, but it does not support combustion as well as pure oxygen because of the high percentage of nitrogen, which will not support combustion. The elements in a mixture can be separated fairly easily because they are not chemically combined. Oxygen in air will support combustion because it is 'free' oxygen. Oxygen in water is not free and therefore will not support combustion. In a mixture there is no fixed proportion of elements composing it. Three different samples of air may contain three different proportions of oxygen, yet each is a specimen of air.

Common Elements

Among over ninety elements some are very rare, but there are a few common ones which should be known, with their symbols:

Barium	(Ba)	a malleable solid (metal)
Calcium	(Ca)	a solid (metal)
Carbon	(C)	a solid (non-metallic)
Hydrogen	(H)	a gas
Iodine	(I)	a solid (non-metallic)
Iron	(Fe)	a solid (metal)
Magnesium	(Mg)	a solid (metal)
Mercury	(Hg)	a liquid (metal)
Nitrogen	(N)	a gas
Oxygen	(O)	a gas
Potassium	(K)	a solid (metal)
Sodium	(Na)	a solid (metal)

Many of these elements are of interest to nurses. *Barium* is a metal. Its salts are opaque to X-rays and can be used for outlining the digestive tract for diagnostic purposes. *Calcium* is the mineral which makes teeth and bones hard. *Carbon* is rare as an element but is present in many compounds and in all living matter, including food. *Hydrogen* is a light, highly inflammable gas and when mixed with oxygen it forms an explosive mixture. *Iodine* is obtained from green food such as vegetables and is present in large quantities in seaweed. It is needed by the body to make the secretion of the thyroid gland. *Iron* is a metal and a small quantity is present in the body. It is necessary for the manufacture of the red cells in the blood and is important for health. Lack of iron is the most common cause of anaemia. *Magnesium* is found in small quantities with calcium in the bones and teeth. *Mercury* is poisonous to the body, but is useful in thermometers as it expands as it becomes warmer. *Nitrogen* is a gas which neither burns nor supports combustion; it is present in all living matter and must be present in our food supply. *Oxygen* is a gas which will not itself burn but which will support combustion; combustion is also called oxidation. *Potassium* is present in all living matter, particularly in plants. In the human body it is present in tissue cells. *Sodium* is also present in all living matter, but particularly in the animal world. In the body sodium is chiefly in the form of sodium chloride. There are approximately 4·5 grams (g) (1 teaspoon) of sodium chloride in 0·5 litre of body fluid (9 g to 1 litre = 0·9 per cent solution).

All compounds and mixtures are made of elements such as these. The body is made of very complicated chemical compounds built up from a small number of elements, so it is necessary for the student of anatomy and physiology to know a little about elements and compounds.

The Structure of Matter

All matter is made up of tiny particles called *molecules* which are so small that they cannot be seen with an ordinary microscope. Molecules are held together only by the attraction of one molecule to another, in the same way that a needle is held to a magnet. It is perhaps rather difficult to accept the fact that an iron bar is made of molecules which do not even adhere to each other but are only held together by mutual attraction. It is, however, an important concept which must be remembered. The force of attraction is called *cohesion* and must be distinguished from adhesion. A molecule is the smallest particle which can exist alone but which still has all the properties of the substance. For example, a molecule of water has all the properties of water which were mentioned earlier.

A molecule can be divided into smaller particles called *atoms*, but an atom cannot exist alone. A molecule of an element consists of atoms of that element, e.g. a molecule of oxygen consists of two atoms of oxygen (written O_2). A molecule of a compound consists of atoms of the different elements present in the compound, e.g. a molecule of water consists of two atoms of hydrogen and one atom of oxygen (written H_2O). A molecule of carbon dioxide consists of one atom of carbon and two atoms of oxygen (CO_2) and a molecule of glucose, which is a simple form of sugar, consists of six atoms of carbon, twelve atoms of hydrogen and six atoms of oxygen ($C_6H_{12}O_6$). These are examples of very simple chemical compounds. In the more complicated compounds which form the human body many elements may be present, and the number of atoms of any one element may run into hundreds. Although all molecules are very small they vary considerably in size and complexity.

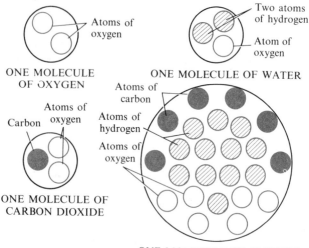

Fig. 2. Diagram to represent molecules of oxygen, water, carbon dioxide and glucose.

Atomic Number and Atomic Weight

The word atom means indivisible and the name was given when scientists thought that the atom could not be divided. Now, however, more is known about the structure of an atom and most people have heard about 'splitting the atom' and its connection with atomic energy. An atom consists of even smaller particles of three different types, **protons** which carry a positive electrical charge, **electrons** which carry a negative electrical charge and **neutrons** which are neutral. The protons and neutrons are massed together to form the nucleus of the atom, while the electrons whirl around the nucleus in one or

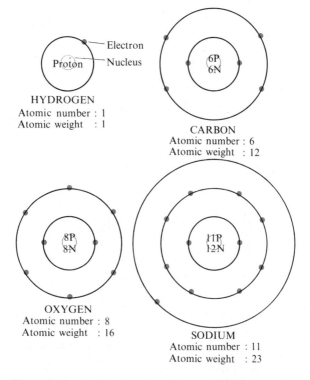

FIG. 3. Diagram of the structures of atoms of hydrogen, carbon, oxygen and sodium. P = proton; N = neutron.

more orbits known as shells. Different elements have different numbers of protons and electrons forming their atoms and the number of protons is equal to the number of electrons. This number is called the **atomic number**. Hydrogen has one proton in its nucleus and one electron circling round it and its atomic number is one. It has no neutron. Carbon has six protons and six neutrons in its nucleus and six electrons, arranged in two shells, circling round it, so its atomic number will be six. Sodium has eleven protons and twelve neutrons in the nucleus, and eleven electrons, in three shells, circling round it, so the atomic number is eleven. These numbers do not have to be memorized because they can be obtained from tables. The presence of neutrons in the nucleus does not affect the atomic number.

The weight of a neutron is roughly equal to the weight of a proton and the number of neutrons added to the number of protons gives the **atomic weight**. The weight of an electron is so small that it is not significant. Hydrogen has an atomic weight of one as it has one proton and one electron but no neutron. Carbon has an atomic weight of twelve as it has six protons and six neutrons in its nucleus. Sodium has an atomic weight of twenty-three as it has eleven protons and twelve neutrons in its nucleus.

Ions and Electrolytes

Atoms do not remain static as electrons may be split off from one atom and gained by another. Atoms are constantly gaining and losing electrons. When a substance is in solution in a fluid and an electron splits from an atom of the substance, an **ion** is formed. Substances which dissociate in this way are known as **electrolytes** because they carry small electric charges. Electrolytes are present in body fluids and are essential to life.

Gas Pressure

Earlier it was stated that one molecule is held to another by mutual attraction. The molecules of a solid are strongly attracted to one another so a solid does not easily alter in size or shape. The molecules of a liquid are less strongly attracted to each other so their positions can be altered more easily and the liquid takes the shape of the container. The molecules of a gas are attracted to each other very little so that when gas is put into a vacuum container the molecules separate from one another and fill the flask. If more gas is introduced the molecules become more densely packed and the gas is said to be under pressure. The pressure of gases is important in their use in the body. Unless oxygen is under sufficient pressure in the lungs it will not be transferred to the blood for transport round the body and the body tissues will suffer from lack of oxygen. On a high mountain the *pressure* of the air is so low that man cannot live without an extra supply of oxygen, though a candle will burn because there is sufficient quantity of air even though the pressure is reduced. In a room where there is reduced quantity of oxygen the candle will not burn, though man can continue to live as the oxygen present is still under pressure.

Solutions and Emulsions

When a lump of sugar is put into a cup of hot water the sugar dissolves and a *solution* of sugar is present in the cup. The molecules of sugar mix with the molecules of water until the sugar is evenly distributed throughout the fluid. The process can be speeded up by stirring, but the even distribution of sugar will occur eventually without any stirring. A measured quantity of fluid will only dissolve a measured quantity of another substance and when the fluid has dissolved as much as possible it is called a *saturated solution*. Fluids can dissolve solids, liquids and gases. Oxygen is dissolved in water, which enables fish to live in it. Heating a fluid drives off the dissolved gas, which appears at the surface as bubbles as the fluid gets hot. Water is the most common solvent and can dissolve a wide variety of substances, but substances which cannot be dissolved in water may be dissolved in other liquids. For example, water will not dissolve oil, but oil will dissolve in spirit. The fact that water is a common solvent makes it important in the body, two-thirds of which is water. Most of this water is in the cells which make up the body, some of it surrounds the cells, which can only live in a salt solution, called saline, and some is in the circulating blood and lymph.

An *emulsion* of oil is different from a solution. If oil is mixed with water the oil will rise to the top when stirring is stopped and will form a layer of pure oil on top of the water. If the oil is mixed with a solution of sodium carbonate (washing soda) an emulsion will be formed. The oil will be broken up into little droplets which will be suspended in the fluid and will not run together again. When allowed to stand the oil will still rise to the top, but the fat droplets will remain separated. Milk is a natural emulsion and the droplets of fat which form the cream can easily be seen under a microscope.

Acids and Alkalis

Substances may be either acid or alkaline in reaction, or

they may be neutral (neither acid nor alkaline). The simplest agent used to detect the reaction of a substance is *litmus* which can be obtained as a fluid, but is conveniently used in the form of litmus paper. Blue litmus paper will turn pink when in contact with acid. Pink litmus paper will turn blue when in contact with an alkali. Neutral substances turn both pink and blue litmus paper into a purple colour. Every living cell must have the correct reaction if it is to survive. The blood and body tissues are slightly alkaline in reaction and vary little throughout life. If too much acid or too much alkali is present in the body illness will occur, and if there is an alteration in the reaction of the blood or body tissues death will follow. It is necessary to know the degree of acidity or alkalinity of certain substances and for this purpose a universal indicator is used. It changes colour from bright red, indicating a strong acid, to purple, which indicates strong alkali. These colour changes are spoken of as the pH of the substance being tested, and the scale runs from pH 1 for the strong acid to pH 14 for the strong alkali. Neutral is pH 7 and the pH of the blood is 7·4.

Pure water is neutral in reaction and as previously stated consists of molecules composed of two atoms of hydrogen and one atom of oxygen. Some of the molecules are broken up into smaller particles, e.g. hydrogen ions (H ions) or hydroxyl ions (OH ions). H ions have a positive electrical charge because the atoms have lost an electron and OH ions a negative electrical charge. In pure water the H ions and the OH ions are equal in number and the reaction is neutral. In some solutions H ions may outnumber OH ions and the reaction of the solution will be acid. If the OH ions outnumber the H ions the reaction will be alkaline. The pH shows the concentration of hydrogen ions (the '*H ion concentration*'), strong acids, e.g. hydrochloric acid, having a high H ion concentration, and strong alkalis, e.g. caustic soda, the highest concentration of hydroxyl ions (OH ions). Some substances, which slow down changes in reaction, are called '*buffer substances*'. In an alkaline solution such as body fluids, buffer substances neutralize acids or any excess of alkalis which may be added to it or produced in it. Examples of buffer substances are sodium, potassium and protein which serve as buffers against acid substances in the body and carbonic acid, lactic acid and fatty acids which serve as buffers against any excess of alkaline bases.

Mineral Salts

A mineral salt is a substance made by the action of acid on a mineral. The mineral is called the *base* of the salt, and the scientific name shows the base and the acid which together form the salt, e.g. sodium chloride is made by the action of hydrochloric acid on sodium. A salt may be acid, alkaline or neutral in reaction. A strong acid acting on a strong alkali forms a neutral salt; a strong acid acting on a weak alkali forms an acid salt; a weak acid acting on a strong alkali forms an alkaline salt. The various salts present in the body differ only slightly in reaction.

Milli-Equivalents

The quantity of each of these salts, e.g. sodium, potassium and magnesium may be measured in milligrams per litre, but is more commonly given in *milli-equivalents* per litre. This term is used to describe the combining power of an atom of a substance compared with the combining power of an atom of hydrogen, which is 1. An atom of any one element will only combine in certain fixed proportions with atoms of other elements. It may be helpful to imagine the atom of each element as having a certain number of strings. Sodium, potassium, hydrogen and chlorine have one string each, oxygen has two strings and carbon has four strings. When the elements combine to form a compound each string must be tied. If hydrogen combines with oxygen two atoms of hydrogen will be needed to one atom of oxygen—this will make H_2O, which is water. One atom of carbon could be combined with two atoms of oxygen to make carbon dioxide (CO_2). This combining power of an atom is called the *valency*. The atoms might be imagined as teams in which the number of men in each team must be equal, although it does not matter what each man weighs.

To find the milli-equivalent value of a substance the number of milligrams per litre is divided by the atomic weight of the element and multiplied by the valency. That is to say

$$\frac{mg/litre}{Atomic\ weight} \times Valency$$

There might be 330 mg of sodium ions in 100 ml of blood.

Number of milligrams	330
Millilitres in 1 litre	1000
Atomic weight of sodium	23
Valency of sodium	1

$$\frac{330 \times 10}{23} \times \text{Valency} = \frac{3300}{23} = 143 \text{ mEq/litre}$$

Diffusion

If two gases of different composition come into contact, intermingling of the gases takes place until the composition of both is the same. For example, the air we breathe out contains more carbon dioxide and less oxygen than the air we breathe in, but the expired air does not float around the room as a portion of different air. The air in the whole room will eventually

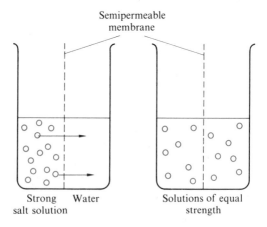

Semipermeable
membrane

Strong   Water
salt solution

Solutions of equal
strength

Fig. 4. Diagram to show the diffusion of liquids.

contain less oxygen and more carbon dioxide overall. This process is known as *diffusion*. The term diffusion is also used to describe the passage of small molecules of acids and salts through a semi-permeable membrane. A membrane may be described as permeable when it allows substances in solution to pass through; as impermeable when no fluid can pass through, and as semi-permeable when water and small molecules in solution can pass through but not larger molecules. If a strong solution of salt is separated from a weaker solution by a semi-permeable membrane the salt molecules will pass through the membrane from the strong solution to the weaker one until both solutions are of equal strength.

Osmosis

If a substance with large molecules, e.g. sugar, is made into a solution and is separated from a weaker sugar solution by a semi-permeable membrane only water will pass through the membrane, from the weak solution to the strong solution, because the molecules are too large to pass. This passage of water across a semi-permeable membrane is called *osmosis*.

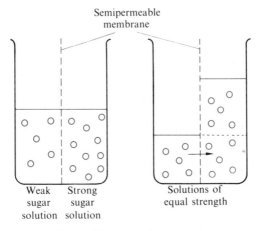

Semipermeable
membrane

Weak sugar solution   Strong sugar solution   Solutions of equal strength

FIG. 5. Diagram to show osmosis.

Specific Gravity

Specific gravity is the weight of a known volume of liquid divided by the weight of an equal volume of pure water.

$$\text{Specific gravity} = \frac{\text{Weight of substance}}{\text{Weight of equal volume of water}}$$

The specific gravity of pure water is expressed as 1·000. The specific gravity of urine ranges between 1·010 and 1·020, and that of blood is about 1·055.

Organic Matter and Inorganic Matter

Organic matter is that which is living or has been alive, e.g. wood or coal. Inorganic matter is that which is not living and never has been alive, e.g. water or iron. Inorganic substances can be built up into organic compounds by living organisms.

2 Characteristics of Living Matter

What is Life?

In the last few years scientists have come very close to an understanding of the mechanisms of life, and this new understanding has shown an underlying unity in all living things. For example, the fundamental molecules and mechanisms in a cabbage are the same as those in a man, and it is becoming possible to give some answers to questions such as 'What is life?' All life is made up from several dozen chemicals put together in similar ways and doing approximately the same job, no matter what organism they come from. And life has another unity. All living matter is made up of small units called cells. Some organisms, such as a bacterium, consist of one single cell; others, like man, consist of many hundreds of millions of cells, all functioning together to make a complete whole.

The Characteristics of Living Matter

All living cells, however simple, have certain characteristics which are always present. These characteristics are:

(1) Activity.
(2) Respiration.
(3) Digestion and absorption of food.
(4) Excretion.
(5) Growth and repair.
(6) Reproduction.
(7) Irritability.

Activity

This is the most striking characteristic of living matter. It is more apparent in the animal world, than in the vegetable world, as the animal must move about in search of food, but even in plants the buds can be seen to break out in the spring, and with a microscope activity becomes as obvious in the plant as in the animal. There can never be any activity without energy. Cars are driven by energy derived from petrol, trains are generally moved by electrical energy, and in living matter the energy is also obtained by burning fuel. The fuel for the human body is the food eaten, particularly carbohydrates and fats. Oxygen is also necessary for the combustion of fuel, and living things obtain their oxygen from the air, or water, in which they live. Combustion of fuel also produces waste products, such as carbon dioxide and water, which must be disposed of.

The combustion of fuel in living matter produces some energy for work and some energy in the form of heat. The body is economical in that it produces more work energy and less heat for each unit of food consumed, but the heat produced by combustion is not altogether wasted, for some heat is needed. The body can remain healthy only within a narrow temperature range of 36° to 37·5°C (97° to 99·5°F) and any heat produced in excess of this must be got rid of if life is to continue.

Respiration

All living matter requires oxygen and gives off carbon dioxide. The oxygen is required for combustion, or oxidation, of food, and carbon dioxide is the waste product of combustion. This process of taking in oxygen and giving off carbon dioxide is called respiration and it is continuous throughout life. The amount of oxygen required and the quantity of carbon dioxide given off vary with the amount of activity taking place. During sleep the human body requires comparatively little oxygen, but it would require much more during strenuous exercise such as playing football or climbing a mountain.

Digestion and Absorption of Food

All living matter requires food. Some food can be absorbed as it is, but most of it must be broken down into substances which have smaller and more simple molecules, before it can be absorbed. The process of breaking down complex foods into simpler substances is termed digestion. It is brought about by enzymes, which are themselves protein substances and which act upon the food, preparing it for absorption.

Excretion

All living matter produces waste products which are of no further use and must be got rid of. If allowed to accumulate, waste products will interfere with the processes of life. In the human body these waste products include carbon dioxide from the lungs, sweat from the skin and urine from the kidneys.

Growth and Repair

Living matter is able to build up new living matter, similar to itself, from food which is taken in. Protein foods include meat, cheese and milk, and these provide the building material for the body. Because the body is continually active its parts are constantly being worn out and they are made good by the building up of new living matter. In childhood, growth and repair are taking place simultaneously. In old age the building up does not keep pace with the breaking down, so weakness begins to appear and will eventually cause death unless illness supervenes. It is this ability to grow and to repair its own tissues, and also the power of reproduction, that makes living matter different from man-made substances.

Reproduction

All living matter can reproduce its own kind. In the simplest forms of life reproduction is a very simple process, involving the splitting of the parent cell into two. In animals, female cells called ova are produced in the ovaries, and sperms, or male cells, are produced in the testes. The sperm cell must fertilize the ovum before reproduction can take place.

Irritability

All living matter is able to respond to a stimulus, and in higher animals this means the power to be aware of the environment and to respond to it. Irritability is more marked in animals, but it can be observed in plants. If a bulb is planted upside down the roots will still grow deeper into the soil and the leaves into the air, or if a plant is placed in a dark corner it will grow tall and thin as it seeks the light. This indicates that even plants are aware of the environment. In animals, irritability allows the awareness of danger, the recognition of that which is useful, e.g. food and water, and the exercise of free will.

All this is what is meant when the physiologist talks about life. Living matter is that which has all the characteristics mentioned above, and all these characteristics are the result of chemical changes. All living matter has the ability to produce protein substances called **enzymes** which in turn control and produce chemical changes within the cell. Enzymes cause other substances to combine with one another, or to split up, though they themselves do not enter into the reaction. For example, complex starches combine with water and split up into simple sugars in digestion; in combustion sugar splits up into carbon dioxide and water. All these changes are brought about by enzymes.

3 The Structure of Living Matter

THE CELL is the unit of living matter, the fundamental building block of life, and all cells are very much alike, regardless of their origin.

Structure of a Cell

All cells are made of a substance called **protoplasm**, which is jelly-like, opaque and colourless, and consists of water and other substances in solution. These other substances include: (a) organic and inorganic salts, (b) glucose, and (c) nitrogenous substances.

The word **cytoplasm** is commonly used to describe the protoplasm which forms the bulk of the cell, the prefix 'cyte' being derived from the Greek word for cell. The cytoplasm is constantly being worn out, broken down and replaced by fresh protoplasm, built up by the cell from the food it takes in. The cytoplasm contains molecules of protein known as ribonucleic acids (RNA) which act as a 'messenger' carrying information out from the nucleus to the cytoplasm.

The **cell membrane** surrounds the cytoplasm and is a semipermeable membrane. The tiny 'pores' in the membrane allow minute molecules to pass into the cell and out of it. The membrane is also thin and elastic and it yields to pressure, so that the cell is able to change its shape.

The **nucleus** is a dense body within the cell, contained within the **nuclear membrane**. The protoplasm inside the nuclear membrane is known as the **nucleoplasm**. The characteristic compounds of the nucleus are deoxyribonucleic acids (DNA), which contain the genetically inherited information required for the maintenance of the cell. The nucleoplasm stores the information necessary for the cell to grow and to divide into two daughter cells. This information is stored in the **genes**, which are strung together to form **chromosomes**.

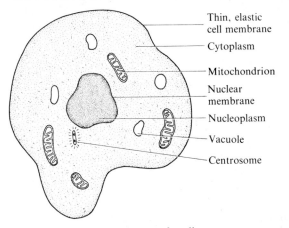

FIG. 6. Diagram of a cell.

Chromosomes are normally only visible, under the microscope, when the cell is about to divide. Genes are composed of DNA.

The **mitochondria** are the power stations of the cell. They are responsible for converting the food ingested by the cell into energy.

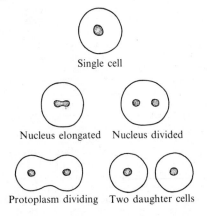

FIG. 7. Diagram illustrating simple fission.

Vacuoles are clear spaces within the cytoplasm, which contain waste materials, or secretions, formed by the cytoplasm.

The **centrosome** is a small rod shaped body near the nucleus and is important in cell division. It is surrounded by a radiating thread-like structure and contains two centrioles.

Cell Reproduction

Simple Fission

The nucleus plays an important part in reproduction. In *simple fission* the nucleus in a single cell becomes elongated, and then divides to form two nuclei in one cell. The cytoplasm then divides between the nuclei to form two daughter cells each with its own nucleus (Fig. 9). These smaller cells will take in food and grow, and when they are full size they will divide to form a total of four cells.

Mitosis

In the more complex forms of life, and this includes the cells in the human body, cell division is a more complicated process, which is called *mitosis*. There are seven stages:

(1) The centrosome divides into two and each moves away from the other, though they are still attached by thread-like structures. This stage is called *prophase*.

(2) The nuclear material forms dark thread-like structures which are the chromosomes. There are forty-six in human cells, though the number differs in other species.

(3) The nuclear membrane disappears and the chromosomes arrange themselves around the centre of the cell. They appear to be attached to the thread-like structure of the centrosomes, which are by now at either end of the cell. These two changes are known as *metaphase*.

(4) The chromosomes divide longitudinally into two equal parts.

(5) The two groups of chromosomes move away to either end of the cell and arrange themselves around the centrosomes. The thread-like structures joining the centrosomes now divide. These two changes are called *anaphase*.

(6) The cell body gets narrower round the centre. The thread-like structures disappear and two nuclear membranes reappear. This is called *telophase*.

(7) The cell divides and the chromosomes disappear into the nucleus. The daughter cells will in turn grow and will reproduce by mitosis.

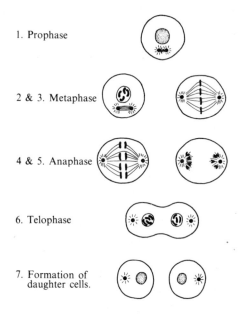

1. Prophase

2 & 3. Metaphase

4 & 5. Anaphase

6. Telophase

7. Formation of
daughter cells.

FIG. 8. Diagram illustrating mitosis.

Meiosis

Reproduction in higher animals, including man, depends on the fusion of a spermatozoon and an ovum. These reproductive cells are also called gametes; fusion of gametes is termed fertilization, and the resulting structure is called a zygote. (When the zygote divides to become a two- and then a multi-celled organism, it is called an embryo.) As previously stated, all human cells contain forty-six chromosomes within the nucleus. The division of a cell, which occurs when gametes are formed, is termed *meiosis* or reduction division. The number of chromosomes must be halved, each gamete receiving one of each pair of chromosomes, but never both, making a total of twenty-three chromosomes in each gamete. When the gametes fuse the resulting zygote will have forty-six chromosomes, the same number as the parent cell. The chromosomes consist of a chain of genes strung together, and it is the genes which pass on the characteristics of the parent cell, so that daughter cells are always similar to parent cells. In this way the physical and intellectual characteristics of the children are derived from the parents, including such characteristics as hair colour, height, degree of intelligence and many others. In any pair of genes one will exert a stronger influence than the other. The stronger gene is termed *dominant* and the weaker one *recessive*. The characteristics which are hereditary depend upon the dominance of the genes.

Sex Determination

One pair of chromosomes from the father and one pair from the mother are the sex chromosomes which will determine the sex of the child. In the female the sex chromosomes are the same and are called XX. In the male they are different and are called XY. One chromosome from each pair will determine the sex of the child. If the child has an X chromosome from the mother and an X chromosome from the father it will be a girl (XX). If the child has an X chromosome from the mother and a Y chromosome from the father it will be a boy (XY).

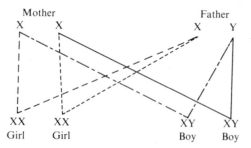

FIG. 9. Diagram to show sex determination.

Unicellular Organisms

These are organisms in which one cell forms a complete individual; examples are bacteria and amoebae. The cell carries out all the characteristic activities of living matter. Movement of the amoeba is performed by the flow of protoplasm within the cell. A projection is pushed out in the direction in which the movement is to be made and the protoplasm streams slowly into the projection until the whole cell occupies a new position. The projection is called a *pseudopodium* or false

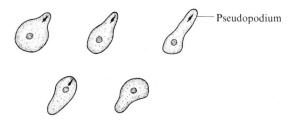

FIG. 10. Diagram to show amoeboid movement.

foot, and this type of movement is called *amoeboid movement*. In the human body it is carried out by the white blood cells. Unicellular organisms are also able to ingest food. Two pseudopodia are put out and the food is surrounded. Enzymes are then poured out and gradually the food is digested and absorbed, until only a small quantity of waste material is left in the vacuole. The cytoplasm then flows back so that the waste material is left behind.

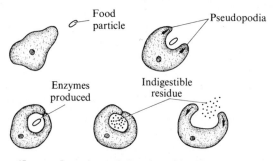

FIG. 11. Ingestion and digestion of food particles.

Multicellular Organisms

Multicellular organisms consist of many cells. Each cell is living and requires food, oxygen, water, suitable temperature and correct pH, but it is dependent on other cells to supply it with the necessities of life in return for the specialized work it carries out. For example, the cells of the lungs take in oxygen, and the cells of the digestive tract absorb food. Each cell develops in a specialized manner so that it can carry out its task satisfactorily. This specialized development is known as the *differentiation* of cells. Groups of specialized cells, developed for a particular purpose, form the *tissues* of the body, e.g. muscle tissue for movement and bone for support.

The multicellular organism in which we are particularly interested is the human body, a complex arrangement of millions of cells, but still having the basic characteristics of all living matter.

4 Systems and Parts of the Body

THE HUMAN body is, as we have seen, an example of a highly developed, multicellular organism. It consists of billions of cells, developed specially to form tissues, e.g. muscle and bone. The body develops from one single cell, the fertilized egg cell or ovum. This *cell* develops into a ball of cells which continue to multiply rapidly and differentiate to form all the various tissues of the body. Each *tissue* has its special function to carry out. These tissues are further grouped together to form organs. An *organ* is a group of tissues arranged in a special manner to carry out a special task, e.g. the stomach, the heart, the kidney, an individual bone or muscle. These organs are again grouped together to form systems. A *system* is a group of organs which together carry out one of the essential functions of the body, e.g. the digestive system is for the purpose of digestion and absorption of food, the respiratory system for the taking in of oxygen and giving off of carbon dioxide. These systems grouped together form the human body.

The body is thus formed of the following nine systems:

(1) The **skeletal system**, to provide a movable framework, which gives support and protection to the softer tissues.

(2) The **muscular system** for movement from place to place.

(3) The **respiratory system** for taking in oxygen and giving off carbon dioxide.

(4) The **digestive system** for the breaking down of food into simple substances which can be absorbed.

(5) The **circulatory system** for transport between part and part; this is essential if there is to be dependence of one tissue on another for the taking in of food and oxygen supplies and for excretion of waste products.

(6) The **excretory system** for the removal of waste matter.

(7) The **reproductive system** for the reproduction of the species.

(8) The **nervous system** for communication between part and part. This is the controlling system which governs the activities of the various organs. Through it awareness of, and response to the environment becomes possible; it enables the activity which we have spoken of as irritability to occur.

(9) The **glandular system**. This is made up of numerous glands which are like the laboratories of the human body producing secretions.

These different systems will be dealt with fully, one by one, in later chapters, but for the better understanding of these chapters it is useful to start with a bird's-eye view of the body as a whole, even though this involves some repetition.

The Systems of the Body

The **skeletal system** consists of many bones forming a rigid, supporting, movable framework. The skeleton permits movements at the joints, the points where two or more bones join, but has no power to move of itself. The **muscular system** is made up of innumerable muscles attached to these bones, to move and hold them in place. It forms the flesh of the body and is responsible for all locomotion (i.e. movement from place to place) and for the power to grasp, hold and reach for things, to move the head, eyes and mouth, to kneel, sit or stand.

The **respiratory system** consists of an air passage leading to the lungs, within which the blood is oxygenated and its carbon dioxide removed. The **digestive system** includes a food canal made up of mouth, pharynx, oesophagus, stomach and small and large intestine. These are hollow, and within them food is acted on by digestive juices, the digestible material is digested and absorbed, while the indigestible residue forms the faeces, which are passed out of the body through the anus.

The **circulatory** or **transport system** includes the blood, the vehicle which carries food, oxygen, waste products and other essentials from organ to organ; the heart, the pump which drives the blood along; and the blood vessels, through which the blood travels. These vessels are of two types: (a) arteries, which carry blood from the heart to the tissues; (b) veins, which carry blood from the tissues to the heart.

The **excretory system** includes: (a) the skin, giving off sweat; (b) the kidneys, excreting urine, which is stored in the bladder and voided when the bladder is full; and (c) the lungs, giving off carbon dioxide.

The **reproductive system**, male and female, produce the

spermatozoa and the ova or egg cells respectively, and are described with the other contents of the pelvis on pp. 36, 37 and 38.

The **nervous system** consists of brain, spinal cord and nerves, which run from the brain and spinal cord to all the tissues of the body. It is similar to the modern telephone system, with the nerves representing the trunk and branch lines, the brain and cord the central exchanges, and the nerve endings in the tissues serving as the telephone receiver with its ear- and mouth-pieces. The nerves consist of nerve fibres, in place of telephone wires; these are of two varieties, one carrying messages to the brain from the tissues which result in sensation, and the other carrying messages from the brain to the tissues which result in movement and activity, just as the telephone cord running from the receiver to the instrument consists of two twisted wires, one to bring the distant speaker's voice to your ear, and one to carry away the sound of your voice to the person to whom you are speaking.

The **glandular system** consists of a large number of glands whose cells are able to take materials from the blood and make new substances from these for some special purpose within the body, just as workers in our chemical industries are able to make carbolic, aspirin, the aniline dyes and countless other products from coal tar. The substances which the glands manufacture are termed secretions; some of them are poured into special organs, where they carry out their work, e.g. the digestive juices are poured into the mouth, stomach and bowel; others pass directly into the blood, by which they are carried throughout the tissues of the body and can stimulate them to activity, e.g. the secretion of the thyroid gland, which lies in the neck, and the suprarenal gland, which lies over the kidney.

The division of the human body into systems is a division according to function or physiology; the body can also be divided up according to anatomy into anatomical parts.

Anatomical Parts of the Body

The anatomical parts of the body are:

    (1) The head.
    (2) The trunk, with the neck.
    (3) The upper limbs.
    (4) The lower limbs.

(See Figs 12 and 13.)

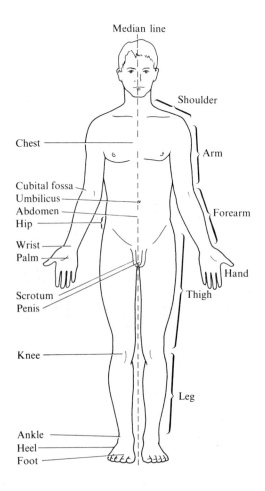

FIG. 12. The anatomical parts of the body (front view).

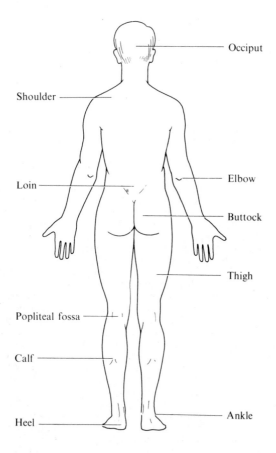

FIG. 13. The anatomical parts of the body (back view).

The *upper limbs* can further be subdivided into:

   (1) The arm or upper arm, from shoulder to elbow.
   (2) The forearm, from elbow to wrist.
   (3) The hand with the digits or fingers and thumb.

The *lower limbs* can similarly be subdivided into:

   (1) The thigh, from hip to knee.
   (2) The leg, from knee to ankle.
   (3) The foot with the digits or toes.

(Note the restricted meaning of the terms arm and leg in anatomy, and be careful to use the terms correctly when describing the positions of bones, blood vessels, muscles, etc.)

The **head and trunk** contain cavities in which lie the vital organs of digestion, respiration, circulation and response to the environment. The walls of the cavities are made of bone and muscle, covered externally by the skin and, under the skin, a layer of fatty tissue called the subcutaneous tissue. This subcutaneous fat forms a reserve of fuel on which the body can draw in time of need, and also serves to keep the body warm, since fat is a bad conductor of heat and so prevents loss of heat from the deeper parts.

The skin, although it lies on the surface in an exposed position, is much more than the paper on the outside of a parcel—in fact, it is one of the vital organs of the body, a point which is often overlooked. It has many important uses. It is the main channel for getting rid of excess heat, which would otherwise raise the temperature of the body and result in death. The blood vessels in the skin are exposed to cooler air passing over the skin surface and heat is lost from the blood by radiation and conduction, and by the evaporation of sweat. This fact was proved in early days when a boy was painted with gold paint from head to foot to take part in a papal procession. The paint checked the action of the skin and the child died of heat stroke. The same problem arose with the waterproof clothing necessary to keep out mustard gas in chemical warfare. It keeps in the heat and at the same time prevents evaporation of sweat, so that it can be worn only for 2 hr at a stretch by an active person. The skin is also a tough protective covering and keeps out bacteria, which cannot make their way through the scaly outer cells of healthy, unbroken skin. (It is for this reason the nurse should cover every break in her skin, however slight.) Again, it is a sensitive coat, rich in nerve endings, which are stimulated by external things, making us aware of heat, cold, pain and touch, as a result of contact with our environment.

The **limbs** are mere outgrowths from the trunk developed in the first place for purposes of movement from place to place. They consist of a central bone, or bones, surrounded by muscles. The bone forms movable joints at either end with the bones of the neighbouring part, e.g. the arm bone forms a joint with the bones of the forearm at the elbow. The muscles are joined to the bones, forming the flesh of the limb and by their contraction give movement to the limbs. The muscles are covered by a layer of subcutaneous fat, as are those of the trunk, and outside this a layer of skin. These have, of course, the same uses as the skin and fat covering the trunk (see Figs 86 and 89).

Through the limbs run blood vessels and nerves, most of them well protected by other tissues for safety. Arteries bring blood rich in food and oxygen to the muscles, bones and other tissues, while veins carry away blood laden with waste products. The nerves carry stimuli from the skin to the brain, through which sensation is experienced, and also convey messages to the muscles, which cause them to contract and produce movement.

Through the centuries man has learnt to stand and walk on two limbs only, using the lower limbs for this purpose. It is a very difficult achievement to learn to balance the body weight on so small a base, and it takes a long time to learn to do it, but once this is achieved the upper limbs are free for other purposes. They are used for lifting, grasping and reaching objects. Man is also able to see to a much wider horizon with the head at a comparatively high level. Because of the different uses to which the upper and lower limbs are put by the human being, there is more difference in structure between them than there is in most animals. The upper limbs are developed to give the greatest possible freedom of movement, while the long digits and the thumb are set at a marked angle to each other to give a wide grasp. Further, the hand contains many small bones, and their freely movable joints make man's hands adaptable for many purposes; there are only three bones from the shoulder to the wrist, but in the hand there are twenty-eight. Through the numerous joints between these bones the hand can make and hold firmly the many instruments man has invented, from the stone axes and arrow heads fashioned in the remote past to the delicate and heavy machinery of the wireless set and the engine-building factories of today. Most of the more powerful muscles which move the joints of the hand lie in the arm, only their tendons crossing the wrist and hand, so that the palm is hollow and the fingers slender, and yet the grip of the hand is strong enough to hold the body weight in climbing and to permit much heavier things to be held during lifting. The lower limbs, on the other hand, are developed to give the greatest possible support and stability combined with spring, steadiness and movement.

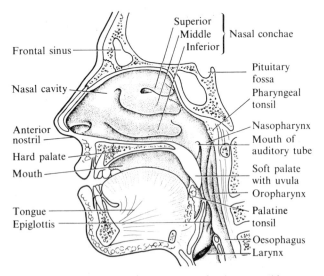

FIG. 14. Sagittal section of the nose, mouth, pharynx and larynx.

The cavities of the head and trunk, together with their contents, should be studied on a model with removable organs, if this is available, and in diagrams.

The **cavities of the head** are:

(1) The *cranium*, containing the brain.
(2) The *orbits*, containing the eyes.
(3) The *nasal cavities*, for the taking in of air.
(4) The *mouth* or *oral cavity*, for the taking in of food.

The Neck

The neck contains the connecting links between the organs of the head and the organs of the trunk, together with bones and muscles to support and move the head. The nasal cavities

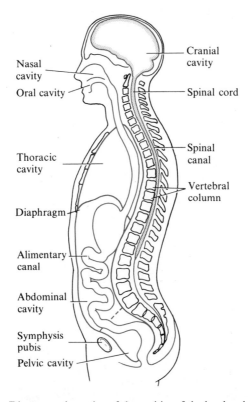

FIG. 15. Diagrammatic section of the cavities of the head and trunk (the unlabelled broken line marks the boundary between the abdomen and pelvic cavity).

and the mouth lead back into one larger cavity, the pharynx, which lies immediately in front of the spinal column. From the pharynx lead the air and food passages, passing to the lungs and stomach respectively. These are:

(1) The **larynx**, containing the vocal cords, lying in the front of the neck and making the prominence known as the 'Adam's apple'; this in turn leads to the **trachea**, which runs

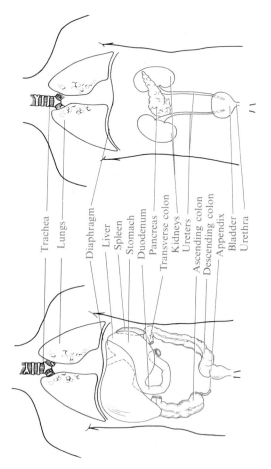

Trachea
Lungs
Diaphragm
Liver
Spleen
Stomach
Duodenum
Pancreas
Transverse colon
Kidneys
Ureters
Ascending colon
Descending colon
Appendix
Bladder
Urethra

Fig. 16. Diagram to show the contents of the cavities of the trunk. The colon is the large intestine. The small intestine is not shown except for the first portion to which the stomach leads.

down the front of the neck and leads into the thorax where it divides into the bronchi: these carry air to the lungs.

(2) The **oesophagus**, which leads from the base of the pharynx and runs down the neck immediately behind the trachea and in front of the spinal column: this carries food to the stomach.

Down the back of the neck runs the spinal column, containing the spinal cord.

The **cavities of the trunk** are:

(1) The *thorax* or thoracic cavity.
(2) The *abdomen* or abdominal cavity.
(3) The *pelvis*, which is continuous with the abdomen.
(4) The *spinal canal*, a posterior cavity inside the vertebral column, communicating with the cranium and containing the spinal cord, which joins the brain.

The **thorax** is the upper cavity of the trunk, and is separated from the abdomen by a dome-shaped sheet of muscle known as the *diaphragm*. It has walls of bone and muscle, and is roughly conical, narrowing towards the neck. The thorax contains:

(1) The *heart* and the large blood vessels leading to and from it.
(2) The *lungs* and the air passages, which bring the air we breathe to them—namely, (a) the *trachea*, which divides into, (b) the two *bronchi*, leading one to each lung.
(3) The *oesophagus*, the passage down which food passes to the stomach.

The **abdomen** is the main lower cavity of the trunk beneath the diaphragm. It has walls of muscle, except where the spine supports it at the back, and where the lower ribs and pelvic bones overlap it. It contains:

(1) The stomach.
(2) The small intestine.
(3) The large intestine.
(4) The liver.
(5) The pancreas.
(6) The spleen.
(7) The kidneys.
(8) The ureters, leading from the kidneys to the bladder; also other smaller structures.

The **pelvis** or pelvic cavity is a basin-shaped cavity surrounded by the bony pelvis or pelvic girdle. The word pelvis means a basin, and the term can be applied either to the bones which form the walls or to the cavity they enclose. It is not shut off from the abdomen by any partition, so that the contents of one cavity can easily slip into the other. It has a floor of muscle called the pelvic floor, which keeps its contents from falling down when we assume the upright position. It contains:

(1) The *bladder*, though, as this fills with the urine excreted by the kidneys, it rises into the abdomen.
(2) The *reproductive organs*.
(3) The *rectum*, the last part of the large intestine.

Coils of the small intestine may also be found in the pelvis according to the size of the bladder and uterus.

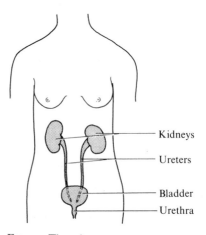

FIG. 17. The urinary system.

The bladder leads to the external skin by the *urethra*. This in the female is short, and opens on to the skin of the vulva, the external genitals (see p. 345). In the male it is much longer, and runs through the penis, opening on to the external surface at its tip. The *rectum* opens on to the external skin in the circular opening called the anus (see p. 235).

The **reproductive organs in the female** are:

(1) The two ovaries, which produce the ova or egg cells.
(2) The uterine tubes, which convey these egg cells to the uterus.

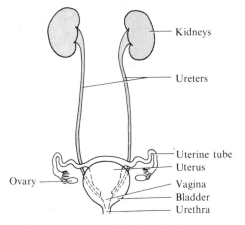

FIG. 18. The female reproductive and urinary organs.

(3) The uterus, in which the fertilized egg cells develop till the child is capable of a separate existence.

(4) The vagina, which leads from the uterus to the external skin.

These lie in the pelvis between the bladder and rectum.

The **male reproductive organs** are:

(1) The testes, which produce the spermatozoa or male reproductive cells.

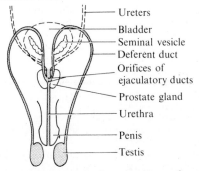

FIG. 19. The male reproductive and urinary organs seen from behind.

(2) The deferent duct, the duct which conveys the male reproductive fluid in which the sperm cells are carried to—
(3) The seminal vesicles, in which it is stored.
(4) The ejaculatory ducts lead from these vesicles to the urethra, the common passage for urine and the male reproductive fluid which contains the spermatozoa.

These organs lie in the pelvis except for the testes which are carried in a pouch of skin, known as the scrotum, at the base of the trunk of the male behind the penis. Here they are exposed to the stimulating effect of cold air, which is necessary to make them function and discharge spermatozoa. The skin at the base of the trunk between vulva and anus in the female, and between scrotum and anus in the male, is known as the perineum.

Definitions of Terms used in Anatomy

The **anatomical position** is the position chosen by the anatomist for purposes of description. It is the upright position with the face towards the observer, the arms hanging at the sides, with the palms facing forwards.

The **middle line** is an imaginary line from the crown of the head to the ground between the feet, used for purposes of measurement and description.

Superior means upper.
Inferior means lower.
Anterior means in front.
Posterior means behind or at the back.
Lateral means farthest from the middle line.
Medial means nearest to the middle line.
Distal means farthest from the head or source.
Proximal means nearest to the head or source.
External means outer.
Internal means inner.
Peripheral means on the outside or periphery.
Ventral means to the front.
Dorsal means at the back.

A **median section** is a section cut down the middle line.

A **transverse section** is a section cut horizontally across the long axis of the body, or of a part.

A **sagittal section** is a vertical section of the skull in an antero-posterior direction, parallel to the median.

A **coronal section** or frontal is a vertical section of the skull at right angles to a sagittal section, i.e. over the top of the cranium from ear to ear, and downwards, dividing the front or anterior part from the back or posterior portion.

Superficial, structures lying near the surface.

Deep, structures lying further in.

5 The Tissues

THE BODY consists of countless cells developed to form various different types of tissue. It originates from a single typical cell, the egg cell or ovum, which is composed of protoplasm and contains a nucleus. After fertilization this cell multiplies and forms a ball of cells, which by the process of differentiation develop into all the various tissues required to form the different organs and parts of the body.

In the very early stages the ball of cells is divided into three layers. The outer layer is called the **ectoderm**, and from it the outer part of the skin develops with its nails, hair follicles and sweat glands and other epithelial tissues including the mucous membrane lining the nose and mouth and the enamel covering the teeth. The nervous system also originates in the ectoderm. The middle layer is called the **mesoderm** and from it muscle, bone and fat develop and some of the internal organs, including parts of the cardiovascular system. The inner layer is called the **entoderm** and from it the lining of most of the alimentary and respiratory tracts develop.

A *tissue* consists of cells and the products of cells specially developed for the carrying out of a special function. In the body there are four main types of tissue.

> Epithelial tissue or epithelium.
> Connective tissue.
> Muscular tissue.
> Nervous tissue.

Epithelial Tissue

Epithelial tissue provides covering and lining membranes for the free surfaces inside and outside the body and is the tissue from which the glands of the body are developed. Epithelia protect underlying tissue from wear and tear, but must continually be renewed as needed. Some cells are specially developed to be able to absorb substances; these linings are only

Table 1
Classification of Tissues

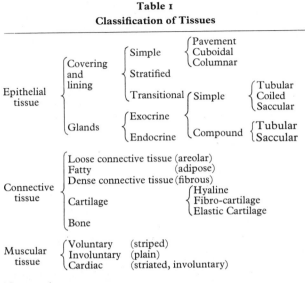

Epithelial tissue	Covering and lining	Simple	Pavement
			Cuboidal
			Columnar
		Stratified	
		Transitional	
	Glands	Exocrine	Simple: Tubular, Coiled, Saccular
		Endocrine	Compound: Tubular, Saccular

Connective tissue	Loose connective tissue (areolar)
	Fatty (adipose)
	Dense connective tissue (fibrous)
	Cartilage: Hyaline, Fibro-cartilage, Elastic Cartilage
	Bone

Muscular tissue	Voluntary (striped)
	Involuntary (plain)
	Cardiac (striated, involuntary)

Nervous tissue

one cell in thickness and often have a specialized surface called a 'brush border'. Some epithelial tissue, particularly glandular tissue, has the ability to secrete substances manufactured within the tissue. Epithelia do not contain blood vessels, the nearest being in the underlying connective tissue, which may be some distance away. The cells are arranged on a 'basement membrane' which plays a part in binding them together.

Covering and Lining Epithelia

Covering epithelia may be classified according to the arrangement and shape of their cells.

Simple epithelium is composed of a single layer of cells attached to a basement membrane; it is very delicate and is found where there is little wear and tear.

(1) *Simple pavement epithelium* is composed of flat cells, which form a smooth lining. These may be found lining the blood vessels and forming the peritoneum.

(2) *Simple cuboidal epithelium* has cells shaped like cubes and is found covering the ovary.

(3) *Simple columnar epithelium* is composed of taller cells, packed onto a basement membrane. It is found where wear and tear is a little greater, for example, lining the stomach and intestines. According to the functions performed columnar epithelium may be modified.

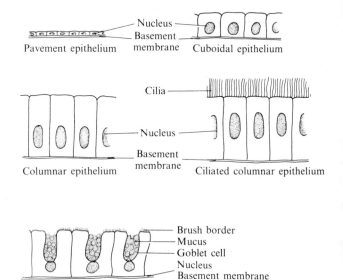

FIG. 20. Diagram showing types of simple epithelium.

Ciliated columnar epithelium has microscopic hair-like processes projecting from the free surface of the cell. The cilia move together with a wave-like motion and mucus and other particles are moved along. This type of tissue is found in the respiratory tract.

Goblet cells are cells which secrete mucus, the mucus collecting in the cell until the cytoplasm becomes distended.

A brush border is found on cells specialized for absorption. These have minute finger-like projections which increase the area through which absorption can occur. They are found in the small intestine.

Stratified epithelium is made up of many layers of cells. The deepest cells, called the *germinal layer*, lie on the basement membrane and are columnar. As they divide, and this occurs frequently, the parent cells are pushed nearer the surface and become flattened. The cells on the surface are rubbed off and are continuously replaced from below. If the surface of the epithelium is dry, as on the skin, the surface cells die because the blood supply is below the basement membrane, and the scaly surface which develops, called *keratin*, makes a water-proof layer. If the surface is moist, as in the mouth, the cells survive until they are rubbed off, so keratin is not formed.

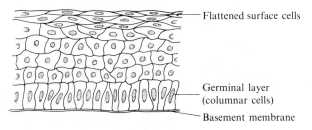

Flattened surface cells

Germinal layer (columnar cells)

Basement membrane

FIG. 21. Diagram illustrating stratified epithelium.

Transitional epithelium is like stratified epithelium but the surface cells, instead of being flattened, are rounded and can spread out when the organ expands. It is found lining organs which must expand and must be water-proof, for example, the bladder.

Glands

Glands develop from epithelial tissues and have the ability to manufacture substances from materials brought by the blood. These substances are called the secretions of the gland. For example, blood contains sodium chloride and the gastric glands can make from this the hydrochloric acid found in the gastric juice, although to obtain hydrochloric acid from sodium chloride in the laboratory is a difficult process. Glands are well supplied with blood vessels which supply the cells with the materials necessary for making their secretions. Glands are of two types.

Exocrine glands pour their external secretions out through a duct. The secretions of many of these glands contain *enzymes*, which are chemical substances produced by the cells of the gland. These enzymes produce chemical changes when they come into contact with specific substances, but do not themselves enter into the reaction.

(1) *Simple glands* have one duct leading from a single secretory unit.

Simple tubular glands are found in the walls of the small intestine, and the stomach.

Simple coiled glands pour sweat onto the surface of the skin.

Simple saccular glands, known as sebaceous glands, secrete a substance called sebum which lubricates hair and skin.

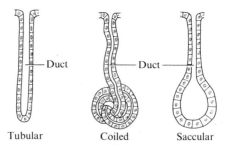

Tubular   Coiled   Saccular

SIMPLE GLANDS

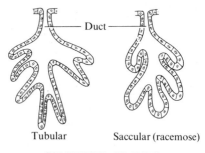

Tubular   Saccular (racemose)

COMPOUND GLANDS

Fig. 22. Types of exocrine glands.

(2) *Compound glands* have several secretory units pouring their secretions into a number of small ducts which unite to form a larger duct.

Compound tubular glands are found in the duodenum.

Compound saccular glands are also called *racemose* glands, and examples are the salivary glands in the mouth.

Endocrine glands pour their internal secretions directly into the blood stream. These secretions are called *hormones*. Examples of endocrine glands are the pituitary gland in the skull and the thyroid gland in the neck.

Connective Tissue

Connective tissue is tissue which supports and binds together all other tissues. There are many varieties of connective tissue, which differ greatly in appearance, though there is similarity in their connective function and in the fact that all originate from primitive cells called *mesenchyme cells*. Connective tissue consists of cells, intercellular substance called *matrix*, and fibres. Matrix and fibres are non-living products of the cells which form the supporting material of the body. The fibres are of two main types.

FIG. 23. Collagenous fibres.

Collagenous fibres originate from cells called *fibroblasts*, which secrete a substance which becomes *collagen*. These coarse fibres occur in wavy bundles and will stretch only a little without tearing.

Elastic fibres are fine branching fibres which are highly elastic.

FIG. 24. Elastic fibres.

In the body a connection sometimes needs to be firm and unyielding and sometimes a degree of elasticity is required. For example, the fibrous layers surrounding organs must be somewhat elastic to allow for swelling when the part is engorged with blood, but the fibrous tendons which join muscle to bone must be inelastic, since if they were elastic the tendon would stretch when the muscle contracted and the bone would not move. There are five main varieties of connective tissue.

Loose Connective Tissue

This type of tissue is also called *areolar* tissue and it consists of a loose network of both collagenous and elastic fibres with small scattered groups of fat cells and some fibroblasts. Some blood vessels and nerves will be found in the tissue but these are not very numerous. Areolar tissue forms a transparent skin, as thin as tissue paper, but very tough, and it is found between and around the organs of the body.

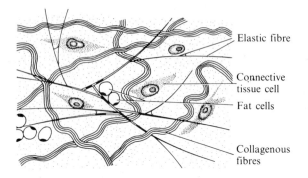

Elastic fibre

Connective tissue cell

Fat cells

Collagenous fibres

FIG. 25. Loose connective tissue.

Fatty Tissue

This is also known as *adipose tissue* and it is similar to areolar tissue but the spaces of the network are filled in with fat cells. Fat cells contain a globule of fat, which pushes the cytoplasm and the nucleus to the edge of the cell. Adipose tissue is useful because it forms a food reserve on which the body can draw in time of need; it helps retain body heat because it is a poor conductor of heat and it protects delicate organs such as the eye and the kidney.

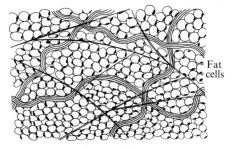

FIG. 26. Fatty or adipose tissue.

Dense Connective Tissue

This tissue is also called *fibrous tissue* and it consists chiefly of bundles of collagenous fibres between which the fibroblasts lie. It is very strong compared to loose connective tissue. The fibres may be arranged regularly, in parallel bundles, as in tendons or ligaments, or irregularly, with the fibres running in different directions, as in the sheath enclosing muscles, which is called *fascia*.

Cartilage

Cartilage consists of cells, called *chondrocytes*, separated by fibres, and it has no blood vessels, so the cells obtain their nourishment by diffusion of fluid through the inter-cellular

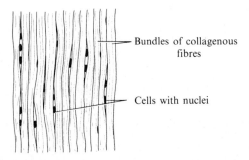

FIG. 27. Diagram of dense connective (fibrous) tissue.

substance. Cartilage is very tough but it is also pliant. There are three types of cartilage.

(1) *Hyaline cartilage* consists of chondrocytes embedded in an apparently structure-less matrix, which is glassy in appearance, and which has very fine collagen fibres running through it. It is found in the trachea and covering the ends of bones at a joint.

(2) *Fibro-cartilage* contains more collagen fibres than hyaline cartilage and is therefore stronger. It is found between bones forming slightly movable joints, for example between the bodies of the vertebrae.

(3) *Elastic cartilage* contains numerous elastic fibres embedded in the matrix and can be found in the auricle of the ear and the epiglottis.

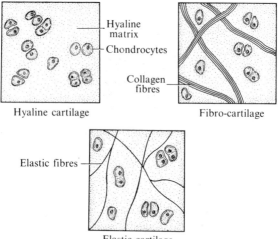

Hyaline cartilage   Fibro-cartilage

Elastic cartilage

FIG. 28. Types of cartilage.

Bone

Bone is a specialized type of cartilage in which the collagen has been impregnated with mineral salts, chiefly calcium. The collagen fibres make the bone tough and the mineral salts make it rigid, so it gives proper support to the soft tissues. The cells between the fibres are called *osteocytes* and bone is richly supplied with blood vessels. The structure of bone will be dealt with more fully in Chapter 6.

Haemopoietic tissue is concerned with the formation of the blood cells and is derived from the primitive mesenchyme cells. Blood may be regarded as connective tissue with the plasma forming the matrix which contains the cells. It will be dealt with more fully in Chapter 12.

Muscular Tissue

Muscular tissue is specialized for contraction and is, therefore, able to produce movement. Wherever there is movement in the body there must be muscular tissue to produce it. Muscle cells are long and thin so that the shortening which occurs during contraction may be as effective as possible. Muscle cells are often called muscle fibres because of their shape. The elastic fibres of connective tissue, if they have been stretched, can recoil and return to their original length, but muscle fibres can, without this preliminary stretching, shorten themselves. There are three types of muscular tissue.

Voluntary or Striped Muscle

Voluntary muscle forms the flesh of the limbs and trunk, by which the skeleton is moved. It consists of long cells, varying in length from a few millimetres in short muscles to from 30 cm or more in long muscles; each cell contains numerous thread-like fibres, called *myofibrils*, which are only from 0·01 to 0·1 mm in width. These myofibrils are regularly striped across in alternate light and dark bands throughout their length, and are so arranged that the dark and light parts come next to one another. Each fibril is enveloped in a sheath of connective tissue called the *sarcolemma*. The fibrils are bound together by connective tissue into bundles, each enclosed in a sheath called the *endomysium*, and these bundles are in turn bound together and enclosed in a sheath called the *perimysium*, to form ultimately the individual muscles, which also have a sheath of fibrous tissue called the *epimysium*. The nucleus is at the edge of the cell. This striped muscle is under the control of the will, so that it is known as voluntary muscle. It contracts strongly when stimulated by a nerve fibre, but tires quickly. For strong contraction much energy is required, so voluntary muscle must have a good blood supply to bring oxygen and nutrients to the cells and to carry away waste products. Capillaries run between individual muscle cells to ensure adequate blood supply.

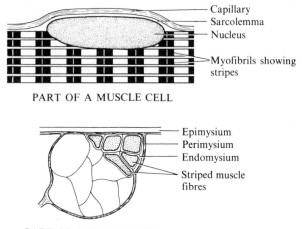

PART OF A MUSCLE CELL

PART OF A VOLUNTARY
MUSCLE

FIG. 29. Sections of voluntary (striped) muscle.

Involuntary or Unstriped Muscle

Involuntary muscle forms the walls of internal organs such as the stomach, bowel, bladder, uterus and blood vessels. It consists of spindle-shaped cells, each of which contains a nucleus. It is also called unstriped muscle. The cells do not show

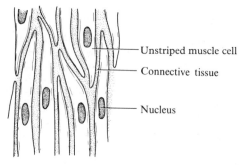

FIG. 30. Section of involuntary (unstriped) muscle.

any stripes and have no sheath, but are bound together by connective tissue to form the walls of the various organs. They are not under the control of the will and act without any conscious effort or knowledge. They contract automatically, but are supplied by autonomic nerves which affect their contractions. This type of muscle is designed for slow contraction over a long period and it does not tire easily.

Cardiac Muscle

Cardiac muscle is both involuntary and irregularly striped. It is found in the heart wall only and is different from any other muscle tissue. It consists of short, cylindrical, branched fibres with centrally placed nuclei. They have no sheath, but are bound together by connective tissue. The cardiac muscle is not under the control of the will, but contracts automatically in a rhythmic manner throughout life, though the rate of these rhythmic contractions is controlled by nerves, which quicken or slow down its action. The fibres branch and join with one another so that impulses can spread from one fibre to another as well as along the length of the muscle.

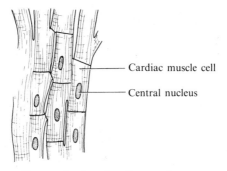

Cardiac muscle cell

Central nucleus

FIG. 31. Section of cardiac muscle.

Nervous Tissue

Nervous tissue is specially designed to receive stimuli from either inside or outside the body, and when stimulated to carry impulses rapidly to other tissues. Nervous tissue consists of nerve cells, called **neurones**, and a special supporting network called *neuroglia*. A nerve cell consists of a large cell body, to which several short processes, called **dendrites**, bring impulses from other cells and tissues. From the cell body there is one long process called the **axon**, which carries impulses away from the cell body.

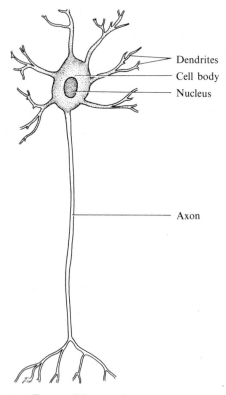

Dendrites

Cell body

Nucleus

Axon

FIG. 32. Diagram of a neurone.

Membranes

The cavities and hollow organs of the body are lined with membranes. These membranes consist of epithelium and secrete lubricating fluids to moisten their smooth, glistening surfaces and prevent friction. Three different types of membrane are found in the body.

(1) *Synovial membrane* secretes a very thick fluid, which is like white of egg in consistency; hence the name, the prefix 'syn' meaning 'like' (from the Greek) and 'ovum' being Latin for an egg. It is chiefly found lining joint cavities to lubricate the movement of the bones on one another. It is a fibrous membrane covered with epithelium. It is also found over bony prominences and between ligaments and bones or tendons and bones. In these sites it forms small sacs called *bursae*, which act as water cushions, facilitating the movement of one part on another. For example, there are bursae around the shoulder, knee and elbow joints. Synovial membrane is also found forming sheaths for tendons which run a long distance to their insertions. For example, the tendons of muscles in the forearm and legs, run across the hand and foot respectively and move the fingers and toes.

(2) *Mucous membrane* secretes a rather thinner fluid called mucus. This membrane is found lining the food canal from mouth to rectum and the air passages from the nose downwards. Thus the cavities which it lines are connected with the external skin. Mucus-secreting tubular glands are also present where much secretion is produced. These are single or branching tubes, lined with secreting cells.

(3) *Serous membrane* consists of flattened cells through which oozes a small quantity of thin watery fluid called serous fluid; this is similar to the fluid which oozes from a blood clot. Serous membrane is found lining the internal cavities, for example the thorax and the abdomen, and covering the organs they contain, providing smooth, glistening, moist surfaces which slide easily over one another when one part moves on, or within, another.

6 Development and Types of Bone

THE skeletal system is made up of about 200 bones, joined together to make a strong, but movable, living framework for the body. It has three main uses. It supports and protects the softer, more delicate tissues and organs, and at the same time provides for movement, since the rigid bones move as levers on one another at the movable joints. It is built up of a special tissue, bone or osseous tissue, which, except for the little ossicles within the middle ear, is normally found only in this one system.

Bone or osseous tissue develops from hyaline cartilage except in the case of the skull bones, which develop from fibrous tissue. In the first place, a mass of cartilage develops of similar shape to the bone which will ultimately be formed, and, in this, mineral salts are deposited, converting it gradually into bone. This depositing of mineral salts is the work of special cells which invade the cartilage from the vascular membrane which covers it (the *perichondrium*). These cells are able to convert the soluble salts of calcium and magnesium in the blood, such as calcium chloride, into insoluble calcium salts, chiefly calcium phosphate, which are deposited in the soft cartilage to harden it. These bone-building cells are known as *osteoblasts*. A second set of cells is also present which bring about the opposite change, and so cause the absorption of any unwanted bone. They can change back the insoluble calcium phosphate into soluble calcium salts, which the blood dissolves and carries away. These bone-absorbing cells are known as *osteoclasts*. Both types of bone cell are active during the period of growth, the bone-builders or osteoblasts producing bone, and the bone-absorbers or osteoclasts removing it to maintain the form and proportions of the bone, e.g. osteoblasts build up bone on the surface of a 'marrow bone,' while bone-absorbing cells eat away bone on the inner side to enlarge the marrow-containing cavity and prevent the bone from becoming too heavy.

This work of bone formation is termed the process of *ossification* and arises from one or more points in each bone which are known as centres of ossification. From these centres the process gradually spreads throughout the whole mass of cartilage or fibrous tissue.

The Composition of Bone

When ossification is complete in the adult, the bone consists of two-thirds mineral matter and one-third animal matter by weight. The *mineral matter* consists chiefly of calcium phosphate, with a small proportion of magnesium salts, and gives the bone its rigidity and hardness. The *animal matter* consists of fibrous material and gives the bone its toughness and resilience.

In young children the proportion of animal matter is much higher, so that the bones are more pliable and break less easily. In old people the bones become more fragile and break with a slight fall, such as follows tripping over a loose carpet.

The Growth and Development of Bone

Since the bones consist largely of calcium and phosphorus, good supplies of these foodstuffs are essential in the diet of the pregnant and nursing mother, the growing child, and where repair of bone is taking place after injury or disease. **Calcium** is especially present in milk and eggs, and in green vegetables. Hard water also contains it, but the inorganic salt present in such water is not so valuable to the body as the organic salts present in plant and animal foods. Where expense make the taking of sufficient quantities of milk and eggs impossible, the calcium of hard water is of some use in eking out the supply. The provision of plentiful supplies of milk for mothers and young children at cheap rates during and since the last war, together with the subsidizing of the cost of eggs by the Government, has greatly lessened the risk of any lack of calcium in the food supplies of infants and children in this country. If there is a deficiency of calcium in the diet of the pregnant woman, calcium from her teeth and bones is mobilized and carried by the blood stream to the growing fetus, so that her teeth readily decay and her bones become soft and deformities occur, which in marked cases are very severe; this condition, known as osteomalacia, is rare in this country, but is common among women in under-developed countries where animal foods are scarce and expensive and pregnancies follow one another very rapidly. **Phosphorus** is present in meat, egg yolk and fish.

Plentiful supplies of the foodstuffs calcium and phosphorus are not, however, all that is necessary for the formation of good bone; supplies of **vitamin D** are also required by the bone-building cells for the creation of bone. This vitamin may be obtained from the fat of fish and animals, provided they have been exposed to the sunlight. It is therefore present in cod-liver oil, milk, cream, butter and fat meat, particularly during the summer months. In the winter New Zealand butter is advantageous in having a higher vitamin content. Margarine does not naturally contain it, but it is now artificially irradiated. Such margarine is cheaper than butter, and is a satisfactory dietic substitute. Since the outbreak of war in 1939 all margarine has been treated in this way, and the process has now been made compulsory.

Cod-liver oil is one of the richest sources of the vitamin, but even here the quantity varies and oil of good quality should be obtained. Cod-liver oil is recommended for pregnant and nursing mothers and for young children, especially during the winter months, but preferably throughout the year. If the oil cannot be tolerated the vitamin can be obtained in various proprietary preparations, but these, though more palatable, are comparatively expensive. They are only necessary when the oil causes vomiting.

The human body can also manufacture the vitamin. Ultra-violet rays of the sun, acting on the ergosterol in the skin, convert it into vitamin D, but as there is little sun during the winter months and we wear clothes and live in houses which cut off the ultraviolet rays, we are largely dependent on the supplies in our food.

Vitamin C is important for the laying down of collagen in bone. It is found in fresh fruit, particularly blackcurrants, oranges and lemons, and in green vegetables and tomatoes.

Growth and development of bone is also affected by both *exercise* and *rest*. Exercise causes an increased blood supply to the muscles of the part, and therefore to the underlying bones; since the blood is the source of the necessary building materials, exercise results in increased growth. Recognition of this fact is responsible for the increased attention to physical exercise in our schools of today. The pull of developing muscles is an important factor in determining the shape of the bones, and posture may also affect shape by altering the stresses on the bones. The importance of rest is interesting. The bones of the child are comparatively elastic so that at the end of a day of standing and running about there is an appreciable loss in height. During rest in the flat position the bones recover their full length. A long night and an hour's rest in the flat position in the middle of the day may therefore affect growth.

A further factor controlling growth is the secretions of certain ductless glands. This will be discussed later in Chapter 21.

Types of Bone Tissue

Bone tissue is of two types.

Compact bone appears to be solid, but when examined under a microscope is found to consist of *Haversian systems*. A Haversian system consists of: (a) A *central canal*, called a Haversian canal, which contains blood vessels, nerves and lymphatics. (b) Plates of bone, called *lamellae*, arranged round the central canal. (c) Between the lamellae are spaces, called *lacunae*, which contain bone cells, called osteocytes, and lymph. (d) Fine channels, called *canaliculi*, run between the lacunae and the central canal carrying lymph to bring food and oxygen to the bone cells. Between the Haversian systems there are tiny circular plates of bone called *interstitial lamellae*.

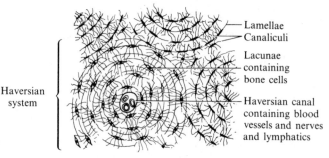

Haversian system

Lamellae
Canaliculi
Lacunae containing bone cells
Haversian canal containing blood vessels and nerves and lymphatics

FIG. 33. The structure of compact bone.

Spongy bone gives a spongy appearance to the naked eye, though it is of course hard like all bone. When examined under a microscope the Haversian canals are seen to be much larger and there are fewer lamellae. The spaces in spongy bone are filled in with *red bone marrow*, which consists of fat and blood cells, and where red blood corpuscles are made.

Types of Bones

There are three types of bones:

    (1) Long bones.
    (2) Flat bones.
    (3) Irregular bones.

Some authorities recognize a fourth type, the short bones, applying the term to the small bones of the wrist and ankle; but these have no special structure to identify them, and are merely so called because, unlike all the other bones of the limbs, they are not long. They are similar to the irregular bones in structure and may be classed with them.

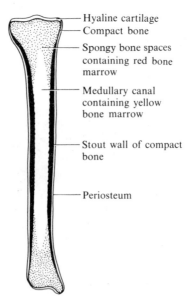

FIG. 34. The structure of a long bone.

Long Bones

A long bone consists of a shaft and two extremities. The shaft has a stout wall of compact tissue surrounding a central canal, called the *medullary cavity*, which contains yellow bone marrow. This bone marrow, like the red bone marrow of spongy bone, consists of fat and blood cells, but is not so rich in blood supply nor in red blood corpuscles. The extremities consist of a mass of spongy bone covered by a thin layer of compact bone. The spongy bone contains red bone marrow, which is of importance in maintaining the number of the red corpuscles in the blood stream. The bone is covered by a tough sheath of fibrous tissue, known as the *periosteum*. This is richly supplied with blood vessels which pass into the bone to nourish it. Three distinct sets of vessels supply the long bones:

(1) Countless minute arteries run into the compact bone to supply the Haversian canals and systems.

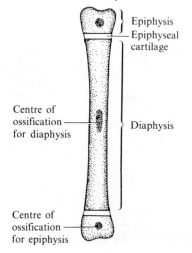

Epiphysis
Epiphyseal cartilage

Centre of ossification for diaphysis

Diaphysis

Centre of ossification for epiphysis

FIG. 35. The development of a long bone (growth in length occurs at the epiphyses).

(2) Numerous larger arteries pierce the compact bone of the extremities to supply the spongy tissue and red bone marrow. The openings by which these vessels enter can readily be seen.

(3) One or possibly two large arteries supply the medullary cavity. These are known as nutrient arteries, and pass through a large opening, known as the nutrient foramen, which runs obliquely through the shaft to the medullary cavity.

These three sets of vessels are linked by their fine branch arteries within the bone. The openings which admit them can readily be seen by examining the surfaces of the different bones.

The *periosteum* serves to nourish the underlying bone through its blood vessels. If it is torn off, the bone beneath dies; on the other hand, if the bone is destroyed by disease, but the periosteum remains healthy, it is able to build up fresh bone. In operations for removal of diseased bone the periosteum is left behind for this reason. The periosteum is also responsible for growth in the thickness of the bone. The surface attached to the bone contains cells, called osteoblasts, which are able to lay down fresh bone. The periosteum is protective in nature, and in some simple cases of fracture remains uninjured, holding the bone fragments together. The periosteum gives attachment to the tendons of muscles.

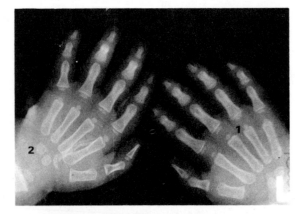

FIG. 36. X-ray of the hand of a 1-year-old infant. The small long bones of the fingers and the thumb have their shafts well formed, but note the gaps (1) between them and the knuckles, where the epiphyses are only beginning to form centres of ossification. Note also that only two of the eight bones in the wrist (2) have begun to ossify.

The periosteum is not present on the joint surfaces of a bone, but is replaced by hyaline cartilage, called articular cartilage. This serves as a buffer to absorb jar and shock between the two bones, and also provides a perfectly smooth surface, so that the movements of the joint occur without friction.

Development of long bones and growth in length.
Long bones develop from three centres of ossification, one in the
shaft and one or more at either extremity (Fig. 35). That in the
shaft develops and is known as the *diaphysis*. Those at the extremities begin to develop after birth, and may not begin to form
till the child is 4 or 5 years old; each extremity starts normally
at a definite time. From these centres, ossification gradually

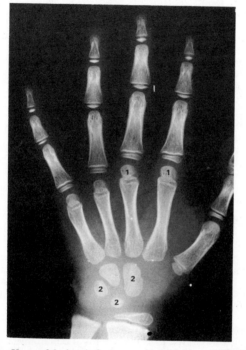

FIG. 37. X-ray of the hand of a 6-year-old child. Note (1) the development of the separate extremities, i.e. the epiphyses, of the long bones
of the hand; and (2) the ossification that has occurred in the wrist or
carpal bones. Only certain bones are marked for comparison.

spreads through the extremity, which is well developed about
the age of 12 years. There is, however, still a line of cartilage
left between the shaft and the extremity. The separate extremity

is known as the *epiphysis* until growth is complete, and the line of cartilage between epiphysis and diaphysis is known as the *epiphyseal cartilage*. It is from this epiphyseal cartilage that growth in length of the long bone occurs. It is the shaft that grows longer, and fresh bone is constantly produced at either end of it by the epiphyseal cartilage. When full growth has been attained, the line of cartilage turns to bone and can no longer be found. This occurs between the ages of 18 and 25 years, occurring at different times in different bones and also at the different ends of one bone, e.g. the lower epiphysis of the arm bone or humerus unites at 18 years, but the upper does not join the shaft till about 2 years later.

Long bones are found in the limbs, where they serve as levers for the movement of the trunk, moving on one another at the joints. Leverage enables movement to be carried out with economical expenditure of energy.

Flat Bones

Flat bones consist of two stout layers of compact bone joined by a layer of spongy bone. They are also covered by perio-steum, from which two sets of blood vessels pass into the bone

FIG. 38. The structure of a flat bone.

to supply the spongy and compact tissue respectively. They are found in the head, trunk and shoulder and pelvic girdles. In the head and trunk they give protection to the delicate organs contained therein. In the shoulder and pelvic girdles they provide a surface for the attachment of the many power-ful muscles required to control the freely movable shoulder and hip joints. The layers of compact bone in the flat bones of the skull are referred to as the outer and inner tables, respec-tively, and the spongy tissue is called the diploë. The outer table is thick and strong so that it is not readily broken. In some regions the spongy bone is absorbed, leaving air-filled cavities called sinuses between the two tables.

Irregular Bones

Irregular bones consist of a mass of spongy bone covered by a thin layer of compact bone. They are covered with periosteum, except on the articular surfaces, which again provides two sets of blood vessels for the supply of the compact and spongy bone. They are found in the head and trunk, e.g. in the vertebrae—and at the ankle and wrist (though these may alternatively be called short bones).

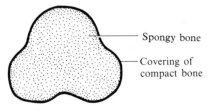

FIG. 39. The structure of an irregular bone.

Surface Irregularities

The surfaces of all bones are very irregular and show numbers of *projections* and *depressions*. These may be divided according to function into:

(1) Articular, which enter into the formation of joints and are smooth.
(2) Non-articular, which give attachment to muscles or ligaments and are rough.

Articular projections are termed:

(1) A head, when round like a sphere or disc.
(2) A condyle, when rounded but oval in outline like the typical knuckle bone.

Articular depressions are termed sockets or sometimes fossae.

Non-articular projections have many terms applied to them according to their nature—namely:

(1) Process, a rough projection for muscle attachment.
(2) Spine, a pointed rough projection.
(3) Tuberosity, a broad rough projection.
(4) Trochanter, a broad rough projection.
(5) Tubercle, a small tuberosity.
(6) Crest, a long rough, narrow projecting surface.

All these rough projections give attachment to muscles, and the stronger the muscle, and the more it is used, the larger and rougher the projection becomes, providing a greater surface for muscle attachment. In a paralysed limb the processes either fail to develop or atrophy, according to age.

Non-articular depressions are termed:

(1) Fossa, a notch in the bone.
(2) Groove, a long, narrow depression.

Other terms applied to bones are:

(1) Foramen, an opening in the bone.
(2) Sinus, a hollow cavity in the bone.

We can now turn to the consideration of the various bones which form the skeleton.

7 Bones of the Head and Trunk

THE *skeleton* is divided into:

(1) The bones of the head.
(2) The bones of the trunk.
(3) The bones of the upper extremities.
(4) The bones of the lower extremities.

The bones of head and trunk form the axial skeleton, the main support of the body, while the bones of the extremities form the limbs and link them to the axial skeleton.

Notice that the term 'extremity' is used here in place of limb. It is not exactly the same thing. The term 'extremity' is used to include both the limb and the girdle by which the limb is attached to the main or axial skeleton. The upper extremity consists of shoulder girdle and upper limb. The lower extremity consists of the pelvic girdle and the lower limb.

The Bones of the Head

The bones of the head are divided into:

(1) The bones of the cranium.
(2) The bones of the face.

The Bones of the Cranium

The *cranium* is a box-like cavity containing and protecting the brain. It has a dome-shaped roof called the skull cap, and its floor is spoken of as the base of the skull or cranium. It consists of the following eight bones:

(1) The frontal bone.
(2) The parietal bones (left and right).
(3) The occipital bone.
(4) The temporal bones (left and right).
(5) The ethmoid.
(6) The sphenoid.

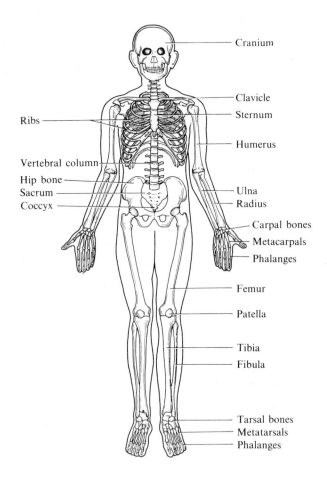

FIG. 40. The skeleton (male).

The *frontal bone* is a large flat bone forming the forehead and the roofs of the orbits. It develops in two parts, which gradually fuse into one bone. It carries two prominences which form the forehead. It contains two cavities called the frontal sinuses, which lie one over each orbit towards the middle line. They contain air, which enters by a small opening leading from the nasal cavities, and are lined with mucous membrane. They may become infected when the nose is infected, as in the common cold. These sinuses give lightness to the bone and resonance to the voice, acting as sounding chambers. Where the portion forming the forehead joins that which forms the roof of the orbit there is a marked ridge, the orbital margin, and above this is a prominence, the superciliary arch which overlies the frontal sinus.

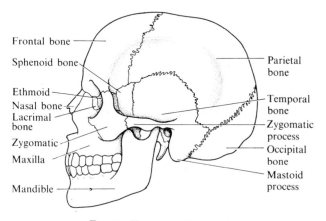

FIG. 41. The bones of the head.

The *parietal bones* are two large flat bones forming the sides and roof of the cranium (*paries* = wall); they articulate with the frontal bone at the vertex of the skull and the occipital bone at the back. On their inner surface, as with all these bones, are small grooves to carry the arteries supplying the brain and larger grooves for the venous sinuses, which take back the deoxygenated blood.

The *occipital bone* is a flat bone forming part of the back and base of the skull (*ob* = against, *caput* = head). It carries a marked prominence, the occipital protuberance, which gives attach-

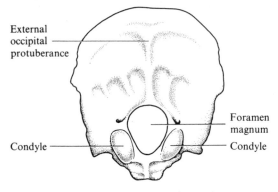

FIG. 42. The occipital bone (from below).

ment to muscles. In it there is a large opening, known as the **foramen magnum**, for the passage of the spinal cord. On either side of the foramen are two condyles, called the occipital condyles, which articulate with the first cervical vertebra, the atlas, and form the joint by which the head nods.

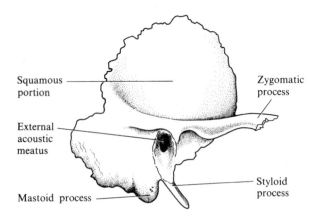

FIG. 43. The temporal bone.

The *temporal bones* are irregular bones, forming the temple at either side of the skull, and also projecting into the base of the skull. They consist of four parts:

(1) A flat plate of bone forms the temple, and is known as the *squamous portion*.

(2) A thick rock-like portion, known as the *petrous portion*, which juts deeply into the base of the skull and contains the internal ear.

(3) The *mastoid portion* behind the ear, which carries a conical projection, the mastoid process, which juts downwards and is easily felt; this gives attachment to a powerful muscle.

(4) The *zygomatic portion*, which carries the zygomatic process; this juts forward in front of the ear and joins the cheek bone to form the zygomatic arch for muscle attachment.

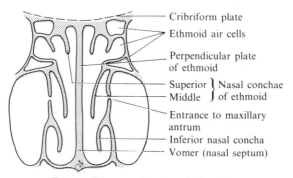

FIG. 44. Diagram of section of ethmoid.

The *ethmoid* is a very light, fragile, irregular bone, and is divided into three parts:

(1) A small horizontal plate which is perforated with many fine openings like a sieve (the cribriform plate). This forms the roof of the nose, and its openings are for the passage of the nerves of smell. It gives the name of the whole bone (*ethmoid* = sieve-like), though it is only a small part of it.

(2) A perpendicular portion which hangs down from the horizontal plate and forms the upper part of the partition or septum in the middle line between the two nasal cavities.

(3) Two spongy portions, one lying on either side. These separate the orbit and nose; they form the outer wall of the

upper part of the nasal cavity and the inner wall of the orbit. They contain a number of air cavities, called the ethmoid air cells, which all communicate with the nose and may be infected from it. These spongy portions carry two processes known as superior and middle nasal conchae, which jut out into the nasal cavities.

The *sphenoid* is an irregular bone, shaped like a wedge (Greek *sphen*, wedge), after which it is named. It lies in the centre of the base of the skull. Its body is in the centre of the base of the skull, and carries a marked depression on its upper surface to accommodate the hypophysis or pituitary gland. This depression is known as the sella turcica or hypophyseal fossa. The body contains the sphenoidal air sinus, communicating with the nasal cavity.

The wings stretch out on either side, and are perforated by many openings for the passage of nerves and blood vessels.

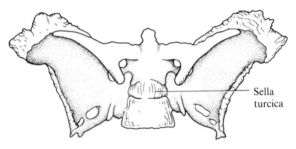

Sella turcica

FIG. 45. The sphenoid bone.

The Bones of the Face

The bones of the face are:

(1) The maxilla.
(2) The mandible.
(3) The zygomatic bones.
(4) The nasal bones.
(5) The vomer.
(6) The inferior nasal conchae.
(7) The lacrimal bones.
(8) The palatine bones.

There is also in the head the hyoid bone at the root of the tongue.

The *maxillae* are two irregular bones which form the upper jaw, meeting one another in the mid-line and being symmetrical. The two bones unite to form one before birth. They carry the upper set of teeth embedded in a projecting piece of bone, called the alveolar process (*alveolus* = a socket cavity). They also form the greater part of the hard palate (which is the roof of the mouth and the floor of the nasal cavities), the outer wall of the nasal cavity, and the floor of the orbit. They are, however, quite light bones, as they contain a large cavity known as the *maxillary sinus*, which has only a thin bony wall surrounding it. This cavity contains air and is lined with mucous membrane. It communicates with the nose from which the air enters it. Again, nasal infection may enter this cavity and cause inflammation.

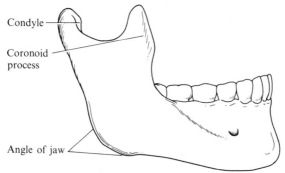

Condyle

Coronoid
process

Angle of jaw

FIG. 46. The mandible.

The *mandible* is an irregular bone, and is the only movable bone in the head. It forms the lower jaw, and carries the lower set of teeth embedded in an alveolar process. It develops in two parts, joined in the middle line by cartilage, which ossifies during childhood. It consists of:

(1) A horizontal portion carrying the teeth.
(2) Two upright portions or rami, which carry: (a) a condyle which articulates with the temporal bone just in front of the ear; and (b) a process, the coronoid process, which gives attachment to muscle.

The upright and horizontal portions meet at either side to form the angles of the jaw. In the bone are openings for the entrance of nerves to supply the teeth.

The *zygomatic bones* are irregular bones forming the prominence of the cheek. They form also part of the floor of the orbit. They join the zygomatic process of the temporal bone to form the zygomatic arch on either side.

The *nasal bones* are two small flat bones forming the bridge of the nose. They unite with one another in the middle line, and articulate with the frontal bone above.

The *vomer* is a flat bone forming the lower part of the septum, or partition, between the two nasal cavities.

The *superior, middle* and *inferior nasal conchae* are small scroll-shaped flat bones, which jut out into the nasal cavities (see p. 192 and Fig. 128).

The *lacrimal bones* are two very small bones on the inner wall of the orbit. They are grooved, and this groove contains the lacrimal sac and nasolacrimal duct, by which the tear or lacrimal fluid, which washes constantly over the front of the eye to cleanse it, is carried down into the nasal cavity. It is only when there is excess of the fluid that it runs down the cheeks as tears.

The *palatine bones* are two irregular bones which form the back of the hard palate and extend up the outer wall of the nasal cavity into the floor of the orbit.

The *hyoid bone* is a small bone shaped like a horseshoe, which lies at the base of the tongue and gives attachment to the tongue muscle. It does not articulate with any other bone of the head or trunk, and lies at the junction of the chin and neck, where it can readily be felt.

The Bones of the Trunk

The bones of the trunk are:

    (1) The sternum or breast bone.
    (2) The ribs.
    (3) The vertebral column.

The *sternum* is a flat bone, shaped roughly like a dagger; it runs down the front of the thorax from the root of the neck and is close under the skin. The upper part is termed the manubrium. The middle portion is named the body. These are joined by a line of cartilage which often does not ossify completely. At this line of junction there is a definite angle between the two parts. The second rib is attached here and it forms a useful landmark. The tip of the sternum consists of cartilage, which never completely ossifies, and is known as the xiphoid process.

The sternum articulates with the clavicles, and the seven pairs of true ribs are joined to it by strips of cartilage. The clavicles join articular surfaces at the upper angles of the bone. Small facets on either side give attachment to the costal cartilages.

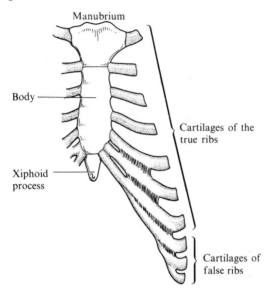

Manubrium

Body

Xiphoid process

Cartilages of the true ribs

Cartilages of false ribs

FIG. 47. The sternum and costal cartilages.

The *ribs* are flat bones. There are twelve pairs, which articulate posteriorly with the twelve thoracic vertebrae. The posterior extremity carries a head which articulates with the bodies of two vertebrae; below this is a narrower neck, and below this a tubercle which is partly rough for muscle attachment, but also carries a smooth facet to articulate with the transverse processes of the vertebrae.

The ribs are divided into:

(1) Seven pairs of *true ribs*; these are attached directly to the sternum by strips of cartilage called the costal cartilages.

(2) Five pairs of *false ribs*. Of these, the first three pairs are not directly attached to the sternum, but are attached to it

indirectly, as their cartilages are joined to those of the rib above; the last two pairs are known as the *floating ribs*, as their cartilages do not join those of the ribs above.

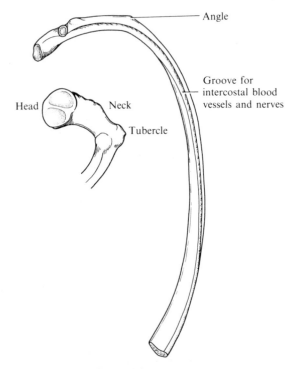

FIG. 48. A typical rib.

The ribs run round the walls of the thorax, sloping downwards towards the front, and also curving downwards between their front and back attachments. They increase in size from above downwards, so that the thoracic cavity is roughly cone-shaped. The first ribs are very small and differ from the others in that they lie practically horizontally. They are grooved for the passage of the large subclavian arteries and veins which run immediately over them to the axillae. The under surface

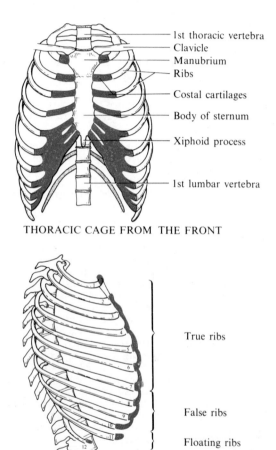

1st thoracic vertebra
Clavicle
Manubrium
Ribs
Costal cartilages
Body of sternum
Xiphoid process
1st lumbar vertebra

THORACIC CAGE FROM THE FRONT

True ribs

False ribs

Floating ribs

THORACIC CAGE FROM THE SIDE

FIG. 49. The thoracic cage.

of each rib is grooved for the passage of intercostal arteries, veins and nerves.

The Vertebral Column

The vertebral column consists of thirty-three irregular bones called vertebrae. Each vertebra consists of:

(1) A disc-shaped *body* lying to the front.

(2) An *arch* of bone jutting out backwards from the body and enclosing a space between body and arch called the vertebral foramen, through which the spinal cord passes.

This arch of bone carries:

(1) Three rough processes for muscle attachment: one spinous process jutting backwards and two transverse processes, one on either side.

(2) Articular processes which carry smooth surfaces to articulate with similar processes on the vertebrae above and below.

The arch carries a notch at either side on the under surface, which forms an opening between bone and bone for the passage of the spinal nerves. The narrow part of the arch above the notch is known as the *pedicle*. The wide part of the arch carrying the spinous process is known as the *lamina*, which forms the back wall of the vertebral column. This is removed in the operation called laminectomy to relieve pressure on the spinal cord after injury or from disease. The vertebrae lie body over body and arch over arch, forming a continuous column which contains a continuous canal. The bodies are joined each to each by thick pads of fibrocartilage called the intervertebral discs. These serve to allow slight movement of bone on bone and yet make very strong joints not liable to dislocate easily; they also absorb jar and shock to prevent its passage to the brain. The arches also articulate with one another posteriorly. The intervertebral discs consist of a ring of fibrocartilage and a softer pulpy centre, called the nucleus pulposus. When the ring is torn the nucleus pulposus can bulge through it and press backwards on to the spinal nerve roots or the spinal cord, causing pain.

The *vertebrae* are divided into five groups:

(1) Seven *cervical* vertebrae.
(2) Twelve *thoracic* vertebrae.
(3) Five *lumbar* vertebrae.
(4) Five *sacral* vertebrae.
(5) Four *coccygeal* vertebrae.

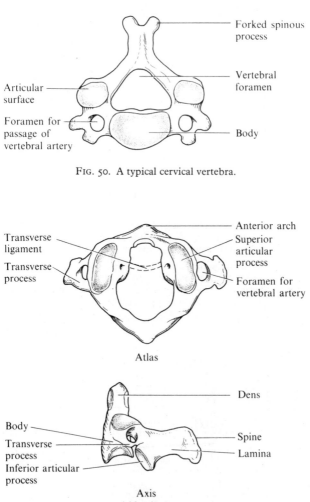

Forked spinous process

Vertebral foramen

Articular surface

Foramen for passage of vertebral artery

Body

FIG. 50. A typical cervical vertebra.

Transverse ligament

Transverse process

Anterior arch

Superior articular process

Foramen for vertebral artery

Atlas

Dens

Body

Transverse process

Inferior articular process

Spine

Lamina

Axis
(side view)

FIG. 51. The first and second cervical vertebrae.

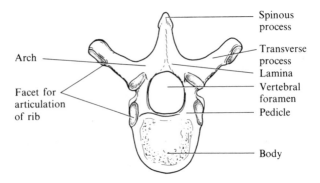

FIG. 52. A typical thoracic vertebra (from above).

The *cervical vertebrae* are the smallest separate vertebrae; they run down the neck, forming a slightly forward curve. They have two distinguishing features:

(1) The spinous process is forked to allow for the passage of a strong ligament which supports the head.

(2) The transverse processes carry openings for the passage of the vertebral arteries to supply the brain.

The first two vertebrae are peculiar, and are known as the *atlas* and the *axis* respectively. The *atlas* is the first vertebra.

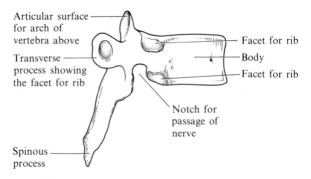

FIG. 53. A typical thoracic vertebra (side view).

It has no body, but is merely a ring of bone. It carries two sockets on its upper surface which articulate with the occipital condyles, forming the joint by which the head nods. The *axis* carries a tooth-shaped projection called the *dens*, which rises from the body and passes through the ring of the atlas and forms a pivot on which the atlas turns, giving the turning movement to the head. A strong ligament runs across between the dens and the spinal cord to hold the process in place and prevent pressure on the spinal cord.

The seventh cervical vertebra has not got a forked spinous process; its process projects very considerably, and can be seen and felt at the base of the neck. It is therefore an important anatomical landmark.

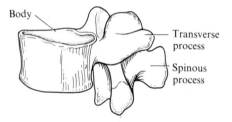

Body

Transverse process

Spinous process

FIG. 54. A lumbar vertebra.

The *thoracic vertebrae* are larger than the cervical and their bodies are roughly heart-shaped. They run down the back of the thorax, forming a backward curve. They are distinguished by two characteristics:

(1) They have long pointed spinous processes which jut downwards.

(2) They carry six facets, three on either side, for articulation with the ribs.

The heads of the ribs lie between the vertebrae, and articulate with one facet on the vertebra above and one facet on the vertebra below—this vertebra being numbered according to the rib which lies above it.

The *lumbar vertebrae* are the largest vertebrae as they have most weight to carry; they run down the back of the abdomen, forming a slight forward curve. Their spinous processes are large and hatchet-shaped, as they give attachment to the powerful muscles supporting the back.

The *sacral vertebrae* are fused together in the adult to form one bone known as the *sacrum*. This runs down the back of the pelvis, forming a backward curve, the upper projecting curve forming the promontory of the sacrum. It is roughly wedge-shaped and articulates at either side with the bone. The spinal canal continues through it, containing a bunch of nerves, the cauda equina; there are openings in the bone leading out from the spinal canal, and by these openings the spinal nerves pass out to the tissues.

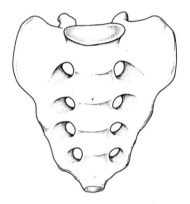

FIG. 55. The sacrum.

The *coccygeal vertebrae* are also fused in the adult to form one bone called the *coccyx*, which articulates with the tip of the sacrum. The bones are very small and imperfect, and form the remnant of the animal tail. The number is usually four, but may be more or less. The joint between sacrum and coccyx allows of slight movement of the coccyx backwards and forwards, which is valuable in childbirth to increase the size of the outlet for the infant.

The vertebral column serves as the main support of the trunk and neck, and gives protection to the spinal cord. It is flexible and gives free movement to the trunk, because of its numerous joints, although each one allows of slight movement only. It serves by its thick pads of cartilage to protect the brain from jars which would result in concussion. It curves give balance to the body. While the fetus is in the uterus there is normally one curve only, a backward curve. After birth, when the child lifts its head, the forward curve in the neck develops, and when the child begins to stand and walk, at about twelve months, the forward curve in the lumbar region develops.

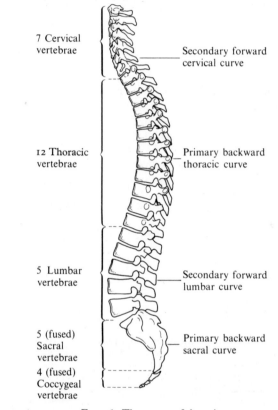

7 Cervical vertebrae

Secondary forward cervical curve

12 Thoracic vertebrae

Primary backward thoracic curve

5 Lumbar vertebrae

Secondary forward lumbar curve

5 (fused) Sacral vertebrae

Primary backward sacral curve

4 (fused) Coccygeal vertebrae

FIG. 56. The curves of the spine.

8 Bones of the Extremities

The Bones of the Upper Extremity

The bones of the upper extremity are:

   (1) The *clavicle* or collar bone   ⎫ forming the shoulder
   (2) The *scapula* or shoulder blade ⎬   girdle.
   (3) The *humerus*, forming the upper arm.
   (4) The *radius* ⎫ forming the forearm.
   (5) The *ulna* ⎬ forming the forearm.
   (6) Eight carpal bones, forming the wrist.
   (7) Five metacarpal bones, forming the palm of the hand.
   (8) The *phalanges*, three in each finger and two in the thumb.

Acromial extremity

Sternal extremity

FIG. 57. The clavicle.

The *clavicle* is a long bone, roughly S-shaped. It articulates with the sternum at its inner or sternal extremity and with the scapula at its outer or acromial extremity. The two extremities are easily distinguishable from one another. The inner extremity is roughly like a pyramid in shape, while the outer is flatter, and in shape and form very similar to the acromion process of the scapula, with which it articulates. The bone lies close under the skin and is easily felt along its whole course; starting from the sternal extremity, it curves first forwards and then backwards. It keeps the scapula in position, and when it is broken the shoulder drops forwards and downwards. It is the only bony link between the bones of the upper limb and the axial skeleton, as the scapula does not articulate with either the ribs or vertebral column. It is a bone which is not found in the skeleton of many four-footed animals, as it only becomes necessary to fix the scapula when the limb is moved outwards from the trunk. The bone is easily broken by falls on the shoulder, as it is compressed between the sternum and the point of impact; it is, in fact, better that it should break than that there should be an injury at the root of the neck, where there are many important structures, or about the actual shoulder joint, where injury would be likely to limit subsequent movement.

The *scapula* is a triangular flat bone; it lies over the ribs at the back of the thorax, but does not articulate with them. It is held in place by muscles which attach it to the ribs and vertebral column. This arrangement gives great freedom of movement to the shoulder girdle, making it possible to reach widely both forwards and backwards and to either side of the body.

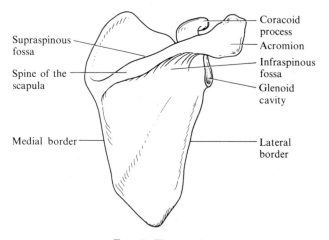

FIG. 58. The scapula.

It is not often broken by falls, as it is embedded in muscle. It has three borders and three angles; the lowest is spoken of as the angle because it is the sharpest, and is easily felt. The front surface is concave to fit over the ribs; the posterior surface is convex, and carries a projecting ridge of bone known as the spine of the scapula, which gives attachment to muscles and forms two depressions or fossae, one above and one below it.

The outer angle carries a shallow socket known as the *glenoid cavity*, which receives the head of the humerus to form the shoulder joint. Above this two processes project:

(1) The *acromion*, the larger of the two, which overlaps the socket and articulates with the clavicle to form the shoulder girdle.

(2) The *coracoid process*, which juts forward and is like a hook.

Both of these can easily be felt. They serve to give attachment to muscles and also help to keep the head of the humerus in place, preventing upward dislocation.

The *humerus* is a long bone and is the largest in the upper limb. Its upper extremity carries a spherical *head* which fits

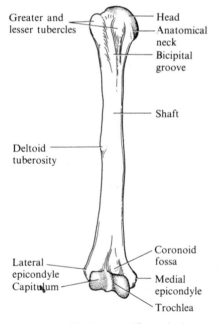

FIG. 59. The humerus (front view).

into the *glenoid cavity*, forming the shoulder joint. The head is attached to the extremity by the anatomical neck, while the extremity is attached to the shaft by the surgical neck, a common site of fracture. The extremity also carries the greater and lesser tubercles for the attachment of muscles. These are divided by a deep groove, which gives passage to one of the tendons of the biceps muscle.

The shaft shows many rough surfaces which give attachment to muscles, the most marked being the deltoid tuberosity on the outer side, which gives insertion to the deltoid muscle. A spiral groove running round the back of the shaft carries the radial nerve, one of the three main nerves of the upper limb.

The lower extremity carries two condyles for articulation with the radius and ulna respectively. Above the condyles on either side are processes called *epicondyles*, which give attachment to muscles and can readily be felt. Above the condyles in the midline of the bone are two depressions, one on the front, and one on the back surface, which receive the processes of the ulna to allow full flexion (bending) and extension (straightening) of the elbow joint. These are named after the processes they receive—the *olecranon fossa* at the back and the *coronoid fossa* in the front. These fossae dip so deeply into the bone that the light can be seen through it here.

The *radius* is a long bone and is the outer bone of the forearm. Its upper extremity is the smaller, and carries a disc-shaped head which articulates with the outer condyle above and the ulna on the inner side. Both these surfaces are smooth and are covered with articular cartilage.

Below the head is the neck, and on the front of the bone there is a process, the *radial tuberosity*, which gives attachment to the biceps muscle. The shaft carries a sharp ridge which faces the ulna, and from this a sheet of fibrous tissue runs to the ulna, joining the two bones throughout their length.

The lower extremity broadens out and carries an articular surface for the carpal bones. It also carries the radial *styloid process*, which can be felt at the base of the thumb.

The *ulna* is a long bone on the inner side of the forearm and slightly bigger than the radius. The upper extremity carries:

(1) Two processes for muscle attachment.
(2) Two sockets for articulation with the humerus and radius respectively.

The processes are known as the *olecranon* and the *coronoid* processes. The olecranon is the larger and forms the point of the elbow. It gives attachment to the triceps muscle and prevents backward bending of the joint. The coronoid process is smaller and projects forwards. These processes help to form the large joint cavity, the trochlear notch, which articulates

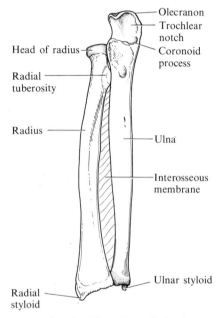

Olecranon
Trochlear notch
Coronoid process
Head of radius
Radial tuberosity
Radius
Ulna
Interosseous membrane
Ulnar styloid
Radial styloid

FIG. 60. The radius and ulna.

with the inner condyle of the humerus very accurately, forming the stable hinge joint of the elbow, which is not readily dislocated. A smaller joint cavity, the radial notch, faces outwards and forms a pivot joint with the head of the radius, which rotates against it. This joint produces the turning movement of the hand. When this rotation of the hand takes place at the elbow the lower end of the radius is carried round the lower end of the ulna, so that it comes to lie on the inner side of the wrist, the shafts of the two bones crossing in the middle of the forearm.

The shaft of the ulna, like that of the radius, carries a sharp ridge for the attachment of the sheet of fibrous tissue, called the interosseous membrane which lies between the two bones.

The lower extremity is much smaller than the upper. It articulates with the radius, but does not join the carpal bones, being separated from them by a pad of cartilage. It carries the ulnar *styloid process*, which is very prominent at the back of the wrist on the inner side.

The styloid processes of radius and ulna give attachment to the strong ligaments of the wrist joint. Both the styloid processes can easily be felt, and normally the radial styloid is slightly lower than the ulna, though this may no longer be so when the radius is broken in an accident, a fact which helps the doctor in making his diagnosis.

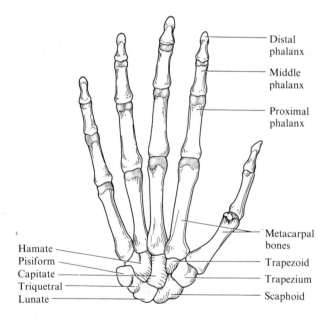

Distal phalanx

Middle phalanx

Proximal phalanx

Metacarpal bones

Hamate
Pisiform
Capitate
Triquetral
Lunate

Trapezoid
Trapezium
Scaphoid

FIG. 61. The hand.

The *carpal bones* are eight small irregular bones arranged in two rows of four. The upper row forms the wrist joint, articulating with the radius, and the lower row articulates with the metacarpal bones. There are gliding joints which allow slight movement between these bones, and since the bones are very small this gives great freedom of movement to the wrist.

The bones are named:

First row, reading from the base of the thumb, inwards:

  (1) The *scaphoid*.
  (2) The *lunate*.
  (3) The *triquetral*.
  (4) The *pisiform*.

Second row, reading in the same direction:

  (1) The *trapezium*.
  (2) The *trapezoid*.
  (3) The *capitate*.
  (4) The *hamate*.

The *metacarpal bones* are small long bones running across the palm of the hand. Their upper extremities articulate with the carpal bones and their lower extremities form the knuckles and articulate with the first phalanx in each digit. The metacarpal bones which articulate with the fingers run almost parallel to each other, but that carrying the thumb lies at a marked angle and can be freely moved both backwards and forwards and to and from the midline of the hand, giving man the ability to grasp and hold all the various tools that he has made.

The *phalanges* are small long bones forming the digits. There are fourteen, three in each finger and two in the thumb.

The Bones of the Lower Extremity

The bones of the lower extremity are:

  (1) The *hip bone*, forming part of the pelvic girdle.
  (2) The *femur*, forming the thigh.
  (3) The *patella* or knee cap.
  (4) The *tibia* ⎫ forming the leg.
  (5) The *fibula* ⎭
  (6) The *tarsal* bones, a group of seven bones, forming the ankle, heel, and instep.
  (7) The *metatarsals*, a group of five bones, forming the sole of the foot.
  (8) The *phalanges*, small long bones, three in each little toe and two in the big toe.

The hip bone is a flat bone helping to form the pelvis or pelvic girdle.

The *pelvis* consists of four bones—the right and left hip bones, the sacrum and the coccyx. It is very different from the shoulder girdle, for whereas the latter moves freely over the chest wall, the bones which form the bony pelvis articulate with the vertebral column and with one another to make a firm girdle, and there is very slight movement only at the various joints. This gives greater stability to the lower limb, but much less freedom of movement.

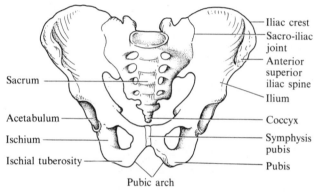

FIG. 62. The male pelvis.

The pelvis is divided into:

    (1) The greater pelvis.
    (2) The lesser pelvis.

The *greater pelvis* is the upper part of the pelvis above the linea terminalis (or pelvic brim). The lesser pelvis is the lower part of the pelvis which actually forms the basin-shaped pelvic cavity and lies below the linea terminalis which is marked by the promontory of the sacrum at the back, the arcuate line of the ilium and the crest of the pubis in front.

The *hip bone* develops in the child in three distinct parts joined by cartilage—namely, the *ilium*, the *ischium*, and the *pubis*.

The *ilium* is the broad flat upper part. It ends above in a broad ridge called the *iliac crest*, which gives attachment to the muscles of the abdominal wall. The crest ends in front in the

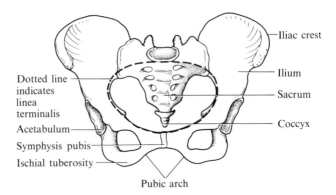

FIG. 63. The female pelvis.

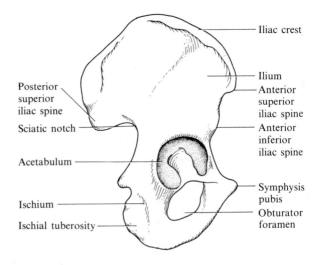

FIG. 64. The hip bone.

anterior superior iliac spine, which is easily felt under the skin, and at the back in the posterior superior iliac spine, which lies under the dimple readily seen on the lower part of the back at either side. The ilium also carries the inferior anterior and posterior spines, which, together with its broad surfaces, provide attachment for the many powerful muscles controlling the hip. At the back there is an articular surface for articulation with the sacrum, and below this a notch called the sciatic notch for the passage of the sciatic nerve.

The *ischium* is the stout lower back portion of the hip bone. It carries the *ischial tuberosity*, which gives attachment to muscles and takes the weight of the body in the sitting position.

The *pubis* is the front lower part of the hip bone, which forms the front wall of the pelvis. The two pubic bones join one another in the middle line, and are united by a thick pad of fibrocartilage; this joint is known as the *symphysis pubis*. The pubis consists of a *body* which enters into the symphysis and carries a ridge, called the pubic crest, and two *branches*, one running up to join the ilium and one down to join the ischium. Between these two branches and the ischium is a large opening called the *obturator foramen*. It lightens the bone and is filled in largely with fibrous tissue.

The three parts of the bone join one another in a deep cup-shaped socket, the *acetabulum*, which receives the head of the femur, forming the hip joint.

The *femur* is a long bone and is the largest and the strongest in the body. Its upper extremity carries a round *head* which articulates with the acetabulum. At its summit is a small depression which gives attachment to the ligament of the head of the femur which runs from the head of the femur to the base of the acetabulum. The head is carried on a long narrow neck which projects at a marked angle from the shaft, only slightly greater than a right angle. This gives a greater range of movement to the limb and in part compensates for the more stable hip-joint. Where the neck joins the shaft there are two marked processes for muscle attachment, the *greater* and *lesser trochanters*. The greater is on the outer side and lies close under the skin, so that it is readily felt.

The shaft is slightly bowed forwards and slopes in towards the knee. It is smooth and rounded except for the posterior surface. Here there is a rough ridge running down the length of the bone, called the linea aspera. This gives attachment to muscles. The ridge is extended above to the greater trochanter in the gluteal line, which gives attachment to one of the gluteal muscles of the buttock.

The lower extremity is broad both transversely and from back to front, and carries two *condyles* which articulate with the tibia, forming the knee joint. The two condyles are separated at the back by a deep notch. In front the two condyles unite with an articular surface for the patella. Above the condyles at the back of the bone there is a smooth surface called the popliteal surface, while on either side are the lateral and medial epicondyles, which give attachment to muscles.

The *patella* is a triangular bone developed in the tendon of the strong extensor muscle which straightens the knee joint

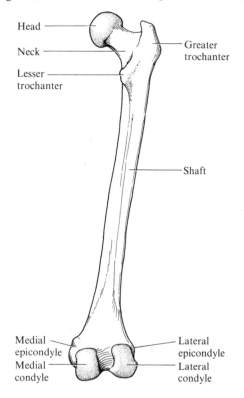

Fig. 65. The femur (back view).

(the quadriceps muscle). Bones which have developed in this way are called *sesamoid* bones. The apex of the patella is downwards, and is joined by the patellar ligament to the tibia. The base is uppermost and gives attachment to the muscle which extends the knee joint. The posterior surface is smooth and covered with articular cartilage. It articulates with the patellar surface on the femur, entering into the knee joint. It allows free movement of the tendon across the joint.

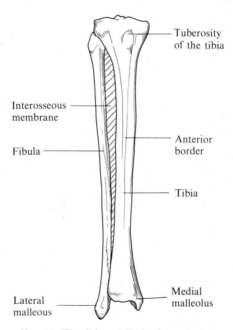

FIG. 66. The tibia and fibula (front view).

The *tibia* is a long bone running down the inner side of the leg, and is much larger than the fibula. Its upper extremity is broad and thick, and carries two shallow sockets which receive the condyles of the femur, forming the knee joint. Between the sockets is a rough area carrying a spine which gives attachment to internal ligaments in the knee joint, the cruciate ligaments.

On the front of the bone is the tuberosity of the tibia which gives attachment to the patellar ligament. There is a small facet on the outer side for articulation with the head of the fibula.

The shaft is roughly triangular in cross section. It has three ridges or borders, the most marked being that on the front of the bone, which forms the *shin*, which can be felt immediately under the skin. Another faces the fibula and gives attachment to a sheet of fibrous tissue (the interosseous membrane) which joins these bones, as the radius and ulna are joined in the forearm. The inner surface of the bone is also subcutaneous and readily felt.

The lower extremity presents an articular surface for the talus below and is attached to the fibula. It carries a process which projects downwards, forming the prominence on the inner side of the ankle joint, the *medial malleolus*. This gives attachment to the ligaments of the joint and serves to prevent dislocation of the ankle joint.

The *fibula* is a slender long bone on the outer side of the leg. The name means 'brooch', and it is like a brooch pin in its relation to the tibia. Its upper extremity, the head, articulates with the tibia just below the knee joint. It does not enter into the formation of the joint. The shaft is slender and ridged, and one of these ridges gives attachment to the sheet of fibrous tissue connecting it with the tibia. The lower extremity forms the bony prominence on the outside of the ankle joint known as the lateral malleolus. This projects below the tibia and articulates with the talus. Like the medial malleolus, it gives attachment to ligaments and helps to prevent outward dislocation of the talus.

The *tarsal bones* are seven in number:

(1) The *talus* or ankle bone.
(2) The *calcaneum* or heel bone.
(3) The *navicular*.
(4) The *cuboid*.
(5), (6) and (7) The *lateral*, *intermediate* and *medial cuneiform* bones.

The *talus* is one of the larger bones of the tarsus and is the only bone which articulates with the tibia and fibula in the ankle joint. This gives greater strength to the joint but less freedom of movement than is to be found in the wrist, the corresponding joint in the upper extremity.

The *calcaneum* lies below the talus and forms the prominence of the heel. It rests on the ground in the upright position and gives insertion to the calf muscles.

The *navicular*, *cuboid* and *cuneiform* bones form the instep.

The *metatarsal bones* are five small long bones articulating with the cuboid and cuneiform bones above and the phalanges below.

The *phalanges* are small long bones, forming the toes. There are fourteen in each foot, two in the great toe and three in each of the other toes.

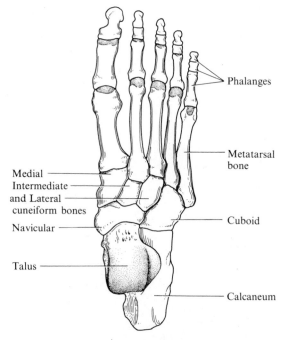

Phalanges

Metatarsal bone

Medial
Intermediate
and Lateral
cuneiform bones

Navicular

Cuboid

Talus

Calcaneum

FIG. 67. The foot.

The Arches of the Foot

The bones of the foot form arches over which the weight of the body is poised. The only bones which touch the ground are the

calcaneum at the back and the front extremities of the metatarsal bones and the phalanges. There are longitudinal arches
on both the inner and outer border of the foot, that on the
inner side being much the higher. There is also a transverse
arch in the tarsal bones.

The arches of the foot give spring to it; they are supported
by strong ligaments in the sole of the foot, and also by the
muscles and the tendons of muscles which cross the sole. These
arches are lost in the condition known as flat foot, in which,
either through overstrain or disease, the muscles become flabby
and stretched, throwing excessive strain on the ligaments,
which give way and allow the bones to get out of place.

9 Joints or Articulations

A JOINT or articulation is formed wherever two or more bones meet and join one another; it is not necessarily a place where one bone moves on another. However, the joints most frequently spoken of are those which do permit movement, for example the shoulders, knees, hips, and elbows. The bones are joined to each other by ligaments, strong cords of fibrous tissue running from bone to bone and strongly attached to the covering periosteum. These ligaments are yielding, but inelastic, and vary in strength and shape according to the work that they have to do and the strain that they have to carry. The ligaments not only allow movement to take place because they are pliable; they also limit movement because of their strength, inelastic nature and rich sensory nerve supply; in this way they protect the joints from excessive strain. Some joints allow no movement of bone on bone, others permit slight movement only, and others a considerable range of movement. As a result, joints are divided into three classes according to the movement which they permit.

(a) Immovable or fibrous joints.
(b) Slightly movable or cartilaginous joints.
(c) Freely movable or synovial joints.

Each type of joint has a different structure, the structure being determined by the movement or lack of movement required at the particular point.

(a) An *immovable* or *fibrous joint* consists of two bones with saw-like edges which dovetail accurately into one another. The bones are at first joined by a line of fibrous tissue, but eventually this becomes ossified, so that bone joins bone and the joint is permanent and does not permit any movement. These joints are found in the skull, and are known as **sutures** from their resemblance to a line of stitching. In the infant at birth there is a definite line of fibrous tissue between bone and

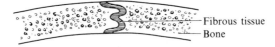

FIG. 68. Diagram of a fibrous joint.

bone which allows the edges of the bones to glide over one another slightly, enabling the head to be moulded to ease its passage through the pelvic canal. At the junction of certain of the cranial bones there are comparatively large expanses of fibrous tissue called *fontanelles*. The largest is the **anterior fontanelle**, which lies at the junction of the two parietal bones and the two parts of the frontal bone at the top of the skull. It is roughly diamond shaped, and normally does not close completely till the child is from 15 to 18 months old. Delay in closure is a sign of rickets.

This area becomes depressed in cases of collapse from dehydration in infancy, and this is a serious sign. A large venous channel runs across the fontanelle from back to front; into this a needle can be passed to obtain a specimen of the blood or to give intravenous injections.

The **posterior fontanelle** lies at the junction of the two parietal bones with the occipital bone at the back of the skull. It is comparatively small and closes shortly after birth. There are also two lateral fontanelles in the base of the skull, but these are not so important to the nurse.

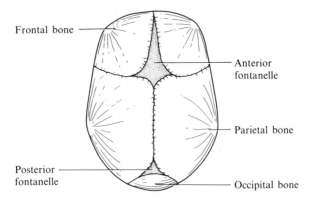

FIG. 69. An infant's head at birth, showing the fontanelles.

(b) A *slightly movable* or *cartilaginous joint* consists of two bones joined by a pad of fibrocartilage, which allows slight movement of bone on bone. Ligaments run from bone to bone around the joint, like a stocking. These joints are of greater strength than the freely movable joints, and are found in the vertebral column and in the symphysis pubis. As a rule there is no synovial membrane, but there may be a partial synovial cavity, as, for example, at the symphysis pubis, where there is a small synovial sac in the centre of the cartilage, between the two bones.

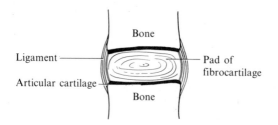

FIG. 70. Diagrammatic section through a slightly movable joint.

(c) A *freely movable* or *synovial joint* consists of two or more bones held together by a *capsule* of ligaments which surrounds the joint cavity. The actual joint surfaces of the bones which come into contact are shaped to fit one another accurately, and are covered with a thick, glistening pad of smooth hyaline cartilage, known as *articular cartilage*. This acts as a buffer between the hard bones, absorbing jar and shock, and also provides a smooth, glistening, frictionless surface, so that the bones can move readily on one another. It has no blood supply and therefore cannot be renewed. It is nourished by the synovial fluid. The joint cavity is lined with *synovial membrane*, which secretes a thick sticky fluid called synovial fluid. This is similar to white of egg in appearance and consistency, and serves to lubricate and nourish the structures forming the joint: the synovial membrane is highly vascular. The capsule of ligaments is loose and elastic, to permit movement, but, in addition, there are a number of ligaments outside it, called extracapsular ligaments, which run from bone to bone. These are so placed that they strengthen the joint without limiting the movements required of it. Certain of these joints also contain additional ligaments within the joint cavity, running from bone to bone. These are termed intracapsular ligaments, and further strengthen the joint by limiting its movements, e.g. the cruciate ligaments of the knee joint.

Some joints are further provided with extra pads of fibrocartilage, which are attached to the joint sockets and serve to deepen them and lessen the risk of dislocation.

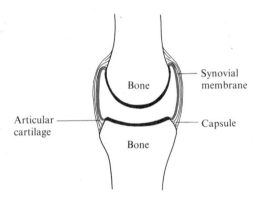

FIG. 71. Diagrammatic section through a freely movable joint.

Types of Synovial Joints

The synovial joints are divided into five different classes:

(1) **Ball-and-socket joints** where a hemispherical head fits into a cup-shaped socket, e.g. shoulder and hip.

(2) **Hinge joints** which allow of movement in one direction only, e.g. elbow, knee, and ankle.

(3) **Double hinge joints** (condylar), which allow movement like a hinge in two directions, e.g. the wrist joint, and the joints between metacarpals or metatarsals and the phalanges.

(4) **Gliding** or **plane joints**, where the bones glide on one another, e.g. between the various carpal and tarsal bones.

(5) **Pivot joints**, where one bone turns on another, e.g. the radius on the ulna at the elbow, and the atlas on the axis, causing the turning movements of hand and head respectively.

Movements of the Various Joints

The movements of which these joints are capable are named:

(1) *Flexion* or bending ⎱ Movements backwards
(2) *Extension* or straightening ⎰ and forwards
(3) *Adduction*, movement to the ⎫
   middle line ⎪ Side-to-side
(4) *Abduction*, movement from the ⎬ movements
   middle line ⎭
(5) *Circumduction*, movements of the limb through a circle, a combination of movements 1 to 4.
(6) *External rotation*, turning a part on its own axis from the middle line.
(7) *Internal rotation*, turning a part on its own axis towards the middle line.
(8) *Supination*, turning the hand palm uppermost.
(9) *Pronation*, turning the hand palm downwards.

These movements may be brought about by a gliding movement of one smooth articular surface on another, by angular movement of bone on bone, the bones acting as levers, or by a rotary or turning movement of one bone on another. Gliding is the simplest form of movement and is sometimes the only movement that the joint permits, as in the joints between the various carpal and tarsal bones and those between the ribs and the vertebrae.

The different types of joints have different kinds of movement. The ball-and-socket joints are most freely movable, allowing flexion and extension, abduction, adduction, circumduction, and external and internal rotation. In the shoulder it must be remembered that the movements flexion and extension are forwards and backwards movements respectively. Hinge joints allow flexion and extension only. There are no hinge joints permitting only abduction and adduction. Double hinge joints allow of flexion, extension, adduction, abduction and circumduction.

In the gliding joints there is slight movement only, increasing the range of movement in all directions, as in the carpal and tarsal bones at the wrist and ankle. Of the pivot joints the atlas, turning on the axis, allows rotation of the head, while the radius, turning on the ulna, permits pronation and supination of the hand.

In the hand and foot, the terms adduction and abduction are used to mean movement to and from the middle line of the part, and not to and from the middle line of the body as a whole. Thus adduction of the thumb brings it to and across the palm of the hand, and adduction of the little finger brings it towards the thumb; in the anatomical position, with the palm facing forward, the little finger is moving from the middle line of the body, but towards the middle line of the hand.

The joints are movable, but the movements are carried out by the various muscles with which the joints are provided. The muscles, in addition to producing movement, run, as do the ligaments, from bone to bone, and help to hold the bones in position and give support to the joint capsule, as long as their normal tone is sustained. When the muscles are paralysed and limp, the looser joints are dislocated comparatively easily, particularly the freely movable shoulder joint; when the muscles are paralysed, but rigid and shrunken, the joint may become completely immovable. This immobility can be prevented by moving the joint through as wide a range of movement as possible to maintain the elasticity of the muscles. Immobility may also result from the joint surfaces becoming adherent to one another, or to the joint capsule as a result of disease or injury affecting the joint itself.

The Joints of the Head

The joints of the head are immovable with one exception, the temporomandibular joint between the lower jaw and the base of the skull. This joint is peculiar; it is a combined hinge and gliding joint, so that movement can take place in all three planes of space—upwards and downwards (elevation and depression), backwards and forwards (regression and protrusion) and from side to side (lateral movement). The joint contains a loose disc of cartilage between the articular cartilages to lessen the danger of jar to the brain and skull from a fall or blow on the chin.

The Joints of the Trunk

There are slightly movable joints between the bodies of the vertebrae. As these joints are very numerous and close together, the column as a whole has free movement. Between ribs and vertebrae there are gliding joints, allowing elevation and depression of the ribs. Between clavicle and sternum is a gliding movable joint containing a pad of fibrocartilage similar to that found in the temporomandibular joint. The joint allows movement of the clavicle upwards and downwards and backwards and forwards.

The Joints of the Upper Extremity

The **shoulder joint** is a ball-and-socket joint, and the most freely movable of the joints in the body. It is formed by the head of the humerus fitting into the small shallow glenoid cavity. The articular surfaces are covered with articular cartilage, and the glenoid cavity is enlarged and deepened by a rim of fibro-cartilage, called the labrum glenoidale, which runs around it. This lessens the risk of dislocation without limiting the movement as much as a larger and deeper bony socket would do. The bones are held together by a loose capsule of ligaments to give the joint its wide range of movement, but the powerful muscles help to keep the bones in position. The long tendon of the biceps muscle serves as an intracapsular ligament. It runs through the bicipital groove between the tuberosities of the humerus into the joint cavity and, since it arises from the scapula immediately above the glenoid cavity, it tends to hold the articular surfaces in position. Movement of the arm above the level of the shoulder joint is due to movement of the scapula over the back of the thorax.

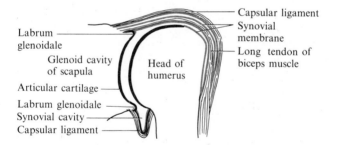

FIG. 72. Diagrammatic section through the shoulder joint.

The **elbow joint** is a complicated one in that it contains the *hinge* joint between the humerus, ulna, and radius and the *pivot* joint between the radius and ulna in one cavity. There is a capsular ligament running between the three bones, and also strong lateral ligaments on either side. A circular ligament, called the annular ligament, also runs round the head of the radius, holding it in the radial notch of the ulna. The lower extremity of the radius also forms a pivot joint with the ulna.

The **wrist joint** is a double hinge joint between the radius and the carpus. It allows flexion, extension, adduction, abduction, and circumduction.

The joints between the carpal bones, and also between carpals and metacarpals, are gliding joints, and, because they are so numerous and the bones so small, they allow a considerable range of movement.

The joints between the lower extremities of the metacarpals and the first phalanx of each digit are of the double hinge or condylar variety, allowing movement in two directions. The joints between the different phalanges of each digit are hinge joints, allowing flexion and extension only.

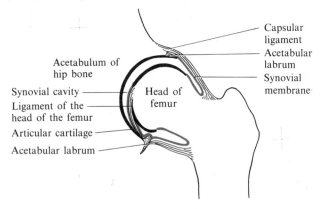

FIG. 73. Diagrammatic section through the hip joint.

The Joints of the Lower Extremity

The **hip joint** is a ball-and-socket joint formed by the head of the femur into the deep cup-shaped acetabulum. The joint surfaces are covered with articular cartilage, and the acetabulum, like the glenoid cavity, is deepened by a rim of fibrocartilage called the acetabular labrum. A strong, round intracapsular ligament (the ligament of the head of the femur) grows out from a depression on the head of the femur and runs to the base of the acetabulum, which is rough and contains a pad of fatty tissue. This ligament serves to limit movement and carries nourishment to the head of the bone. The joint has a strong capsular ligament which also limits movement to some extent in comparison to the corresponding joint in the upper extremity—the shoulder; however, it increases the stability of this joint, which carries the body weight in the standing position, and is under great strain when we land heavily on our feet, in jumping and falling from a height. There are also strong extracapsular ligaments, one which is particularly strong crossing the front of the joint; this helps to maintain the erect position, preventing the body from falling backwards at the hip. Powerful muscles also support the joint.

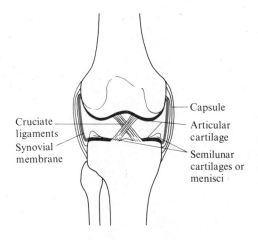

Cruciate ligaments
Synovial membrane
Capsule
Articular cartilage
Semilunar cartilages or menisci

FIG. 74. Diagrammatic section through the knee joint.

The **knee joint** is a hinge joint formed by the condyles of the femur and the tibia and the posterior surface of the patella. It allows flexion and extension and slight side-to-side movement when the knee is flexed. The joint has a large joint capsule into which the patella enters, and is strengthened by both extracapsular and intracapsular ligaments. The extracapsular ligaments are the medial and lateral ligaments, running between the epicondyles of the femur and the fibula and tibia, respectively. The intracapsular ligaments are known as the cruciate ligaments. They run from the tibia to the intercondylar notch of the femur, crossing each other within the joint. They limit movement and therefore strengthen the joint.

The joint is further strengthened by the menisci; these are

two crescentic wedge-shaped pieces of compressed fibrous tissue fixed to the borders of the sockets on the upper surface of the tibia. They deepen the sockets, the thick edge of the wedge being along the outer border of each socket. The cartilages may get out of place or torn as a result of injury to the ligaments which fix them to the tibia. This may fix the joint in a bent position so that it cannot be straightened.

The proximal joint between the tibia and the fibula is of the gliding variety, the distal extremities forming a fixed fibrous joint.

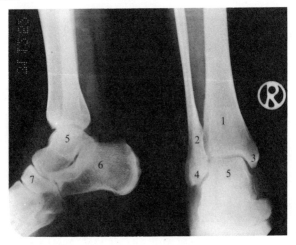

FIG. 75. X-rays of the right ankle joint, showing (left) lateral view, and (right) antero-posterior view. (1) The tibia, (2) the fibula, (3) the medial malleolus, (4) the lateral malleolus, (5) the talus, (6) the calcaneum, and (7) the cuboid.

The **ankle joint** is a hinge joint formed by the tibia, fibula and talus. The movements are flexion and extension, but are spoken of, as a rule, as dorsiflexion, raising the foot, and plantar flexion, raising the heel.

The joints between the various tarsal bones and between the tarsus and metatarsus are gliding joints. They allow flexion and extension, slight abduction and adduction, and inversion, turning the foot sole inwards, and eversion, turning the foot sole outwards.

The joints between the front extremities of the metatarsals and the phalanges are double hinge joints permitting flexion, extension, abduction, and adduction; as these joints are little used if tight shoes are worn, their action should be encouraged by exercises and the wearing of sensible shoes. The joints between the phalanges in each digit are hinge joints, allowing flexion and extension only.

10 Structure and Action of Muscle

THE muscular system consists of a large number of muscles through which the movements of the body are carried out. These muscles are attached to the bones for the most part and produce movement of bone on bone at the many joints in the skeleton. There are exceptions; for example in the head, many of the muscles are attached to the skin at one or both extremities and cause movements of this, as in the acts of frowning and raising the eyebrows. The muscles form the flesh of the body and make up a fifth of its total weight.

The Structure of Muscles

The skeletal muscles are built up of striped special tissue which is known as *voluntary* or muscular tissue. This tissue consists of fine, thread-like muscle fibres, striped across in alternating light and dark bands (see p. 49). Each fibre consists of soft contractile material which is enclosed in a connective tissue sheath, the *sarcolemma* (the Greek words *sarx* = flesh; *lemma* = husk), and is joined to its neighbours by loose connective tissue to form bundles similarly enclosed in sheaths. These bundles, bound together end to end and side to side, form larger and larger bundles, which ultimately form the individual muscle, made up of many thousands of fibres. The muscle is enclosed in a fibrous sheath called the epimysium. Each muscle is again bound by loose connective tissue to the neighbouring muscles or other tissues which make contact with it, and groups of muscles in various regions are enclosed in a common sheath which is known as fascia, for example, in the arm and thigh all the various muscles of the part are enclosed in a single sheath which lies immediately under the subcutaneous fat and is joined to it by loose connective tissue. This sheath is particularly strong in the thigh, where it is called the *fascia lata* (*lata* is the Latin word for broad).

The connective tissue which runs through the muscles forms a supporting framework to the muscular tissue and carries the blood and nerve supply to the muscle fibres. The muscle fibres are able to contract or shorten and so produce movement. Since they are active tissues they are well supplied with blood vessels; arteries bring them food for fuel and repair and oxygen for combustion of the fuel, while veins carry away the waste products of their activities such as carbon dioxide. Blood vessels and nerves enter and leave the muscle at the hilum. The nerve supply is both motor and sensory. The motor nerve supply brings stimuli to the muscle fibre, causing

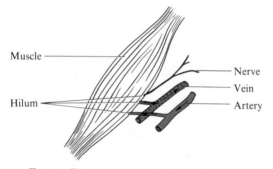

FIG. 76. Diagram to show the hilum of a muscle.

it to contract. It produces a chemical change in the muscle which results in contraction of the muscle fibres. The sensory nerve supply gives us 'muscle sense,' not a very acute sensation, but sufficient to make us aware of contraction and relaxation in the muscles. We are not consciously aware of this sensation normally, but it is at once obvious when we begin to think about it.

Normally the muscles of the body are in a state of slight contraction, known as muscle tone, ready to go into contraction immediately they are stimulated by the nerves. This muscle tone depends partly on exposure of the skin to cold air. The colder the weather the greater the muscle tone, and vice versa. Hence the bracing effect of a cold day or a cold bath and the relaxing effect of great heat.

The Shape of the Muscles

The muscles of the body are for the most part *spindle-shaped*, with a thick belly of muscular tissue which tapers away at either end to form a strong cord of fibrous tissue called the tendon; which attaches the muscle to a bone at each end. These typical muscles are found chiefly in the limbs. In the trunk and head many of the muscles are sheet-like, forming the walls of cavities. These sheet-like muscles sometimes arise direct from bone without intervening tendon; this is termed a fleshy attachment, although, in fact, there is always a small tendinous portion attaching the muscle fibres to the bone. In other cases they are attached either to the bone or to another muscle by sheets of fibrous tissue, or flattened tendons, called *aponeuroses*.

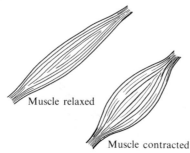

Muscle relaxed

Muscle contracted

FIG. 77. A typical voluntary muscle showing contraction.

The Action of the Muscles

When a muscle contracts, one end normally remains stationary while the other end is drawn towards it. The end which remains stationary is called the *origin* and that which moves is called the *insertion*. Both origin and insertion are usually points of attachment to bone. Every muscle has its definite origin and insertion, and normally the bone of insertion is pulled upon when the muscle contracts and drawn towards the origin, one bone moving on the other at the joint. On the other hand, it is possible to use a muscle, as it were, the wrong way round; the bone into which the muscle is inserted may be temporarily fixed by the action of other muscles, so that it remains stationary and the pull is exerted on the point of origin; for example, a large muscle (the psoas) runs across the front of the groin, originating in the bodies of the five lumbar vertebrae and being inserted into the lesser trochanter of the femur. Normally this muscle pulls on the femur across the front of the groin and produces flexion of the hip, bending the thigh on the trunk. If, however, the femur is temporarily fixed by the pull of other muscles upon it, when the psoas muscle contracts it will pull on the lumbar vertebrae and bend the trunk forwards on the thigh. This arrangement economizes in the number of muscles required. Further economy is obtained by the clever placing of many muscles, which enables one muscle to carry out more than one action. Muscles must cross the joint which they move. Some of them cross two joints, and therefore produce movement in more than one joint, e.g. the biceps crosses both the elbow and the shoulder, causing flexion of both joints.

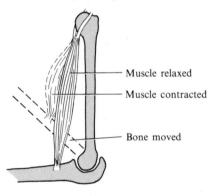

Muscle relaxed

Muscle contracted

Bone moved

FIG. 78. Diagram showing bone movement as muscle contracts.

Muscles produce action by contraction only. They contract and pull; they never push. When the contraction occurs the muscle may become shorter, thicker and harder and produce movement at a joint; or it may contract without shortening, merely holding the joint fixed more or less firmly according to the degree of the contraction; also muscles can contract as they lengthen, so that there is a check to balance the action of opposing muscles. When the contraction passes off the muscle becomes soft, but does not of itself lengthen. It is, however, elastic, and can be stretched by the contraction of another muscle on the opposite side of the joint. Muscles are therefore always grouped round the joints in pairs, the two muscles of each pair having opposing actions and being known as antagonistic muscles.

The muscles are named according to the movement they produce—flexors, extensors, adductors, abductors and so on. Every flexor is opposed by an extensor, a supinator by a pronator, an internal rotator by an external rotator.

Muscles seldom work alone. Even the simplest movement usually involves the action of many muscles, for example, picking up a pencil requires movements of the fingers and thumb, the wrist, elbow, and possibly of the shoulder and trunk, as the body leans forward to reach it. Each muscle involved must be contracted just sufficiently, and not only must each muscle contract, but the opposing muscle must relax to permit the movement to take place. This concerted action of many muscles is known as muscle *co-ordination*. In any new action which we undertake involving a new muscular co-ordination we have great difficulty until the new co-ordination has been learnt, e.g. knitting, swimming, dancing, driving a car. Once the new co-ordination has been acquired, it becomes an easy matter and we can carry it out without mental effort.

The act of standing is a very good example of this. It involves the use of muscles to control the ankle, knee, hip, vertebral column, and head; in fact, muscles from the sole of the foot to the crown of the head. The co-ordination is very difficult to learn and takes a long period of time, but once the power to balance the body on two feet has been acquired by the co-ordinated use of the various muscles of the body, we can stand easily without mental effort.

Contraction of Muscle

Muscle is composed of:

75 per cent water
20 per cent protein
5 per cent mineral salts, glycogen and fat.

Muscle contraction occurs as a result of nerve impulses. In order for muscle fibres to contract energy is required and this is obtained from the oxidation of food, particularly carbohydrates. During digestion carbohydrates are broken down to a simple sugar called glucose. The glucose which is not required immediately by the body is converted to glycogen and is stored in the liver and in the muscles. Muscle glycogen provides the source of heat and energy for muscular activity. During the oxidation of glycogen to carbon dioxide and water a compound is formed which is rich in energy. This compound is called adenotriphosphate (ATP). When it is necessary for muscular contraction the energy from ATP can be released as the compound changes to adenodiphosphate (ADP). During the oxidation of glycogen pyruvic acid is formed. If plenty of oxygen is available, as it usually is during ordinary movement, pyruvic acid is broken down to carbon dioxide and water, and during the process energy is released, which is used to make more ATP. If insufficient oxygen is available the pyruvic acid is converted to lactic acid, which accumulates and produces muscle fatigue.

During violent exercise more oxygen is brought to the muscles, but even so not enough oxygen reaches the muscle cells, particularly at the beginning of the effort. Lactic acid accumulates and diffuses into the tissue fluid and the blood. The presence of lactic acid in the blood stimulates the respiratory centre and the rate and depth of respiration are increased. This continues, even after the exercise is over, until sufficient oxygen has been taken in to allow the cells of the muscles and the liver to oxidize the lactic acid completely, or to convert it to glycogen. This extra oxygen needed to remove the accumulated lactic acid is called the 'oxygen debt', which must be repaid after exercise is completed.

Table

Changes during muscle contraction

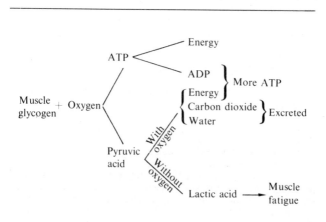

11 The Chief Muscles of the Body

The Muscles of the Head

The muscles of the head are divided into two groups according to their functions, namely: (a) the muscles of expression, and (b) the muscles of mastication.

The **muscles of expression** are attached to the skin rather than to the bone, so that they move the skin and change the facial appearance. Circular muscles, called orbicularis oculi and orbicularis oris, surround the eyes and mouth respectively, closing them. Small muscles raise and lower the eyebrows and upper lids, raise and lower the angles of the mouth and dilate the nostrils, causing a look of surprise, worry, happiness or sorrow. Small muscles also move the eyeballs in the orbits (see p. 319) both to direct the eyes to objects for sight and also to change the expression.

The **muscles of mastication** move the lower jaw up and down in biting, and also both from side to side and backwards and forwards in chewing. They are the *masseter* running to the angle of the jaw from the zygomatic arch, the *temporalis muscle*, lying over the temporal bone and inserted into the lower jaw, and the smaller muscles which also run from the skull to the lower jaw. These are the muscles which 'lock the jaw' in tetanus.

The Muscles of the Neck

The neck contains two large muscles, the sternocleidomastoid and the trapezius.

The *sternocleidomastoid* lies in the front of the neck, running from the sternum and clavicle to the mastoid process and the surface of the temporal bone behind it. When the muscle on one side contracts it draws the head towards the shoulder. When both are used together they flex the neck.

The *trapezius* lies over the back of the neck and chest and is

Table 3
Muscles of the Neck

Name	Position	Origin	Insertion	Action
The sternocleido-mastoid	The front of the neck	Sternum and clavicle	Mastoid process	Used separately, turn the head to the side: used together, flex the neck
The trapezius	The back of the neck and chest	The occiput and the spines of the thoracic vertebrae	The spine and the scapula and the clavicle	Draws the scapula back, bracing the shoulders; the upper part raises the shoulder, the lower part lowers it; the upper part can also be used to pull on the occiput, extending the neck

roughly triangular in shape, with the base joining the spine down the back of the neck and chest, from the occiput, to which it is also attached, downwards; the angle is inserted into the scapula and clavicle, over the top and back of the shoulder. It draws the shoulders back when used as a whole, and also draws the scapula up and down, when the upper and lower portions are used separately (see Table 3).

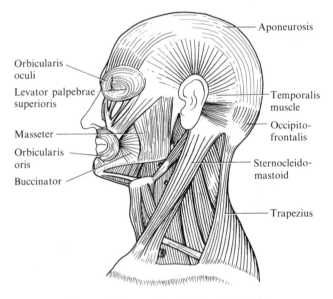

FIG. 79. The muscles of the head and neck.

The Muscles of the Trunk

The chief muscles of the trunk can be grouped according to their function into (a) muscles moving the shoulder, (b) muscles of respiration, (c) those forming the abdominal wall, (d) those moving the hip, (e) those moving the spine, and (f) the muscles of the pelvic floor.

Muscles Moving the Shoulder

The chief muscles moving the shoulder are the powerful muscles covering the back and front of the chest. They include the pectoralis major, the trapezius (see p. 115), the latissimus dorsi and the serratus anterior. The *pectoralis* covers the front of the chest, running from the sternum out to the humerus. The *latissimus dorsi* covers the back of the chest and abdomen, running from the lumbar vertebrae and iliac crest out to the humerus. These muscles form the muscle in front of and behind the armpit. The *serratus anterior* runs round the side wall of the thorax from the ribs in front to the vertebral edges of the scapula under which it passes. (For actions and details see Table 4.)

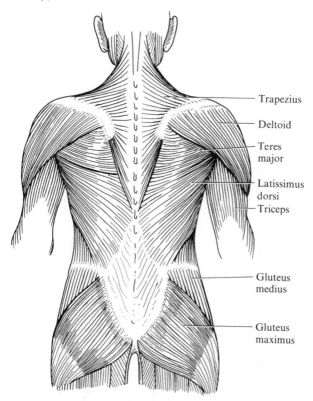

Trapezius

Deltoid

Teres major

Latissimus dorsi

Triceps

Gluteus medius

Gluteus maximus

FIG. 80. The muscles of the back.

Table 4

Trunk Muscles moving the Shoulder

Name	Position	Origin	Insertion	Action
Pectoralis major	Front of chest	Sternum and clavicle	Humerus (outer edge of bicipital groove)	Adduction of the shoulder, drawing the arm across the front of the thorax. Also internal rotation of the shoulder
Latissimus dorsi	Crosses the back from lumbar region to the shoulder	Lumbar vertebrae, lower thoracic vertebrae, and back of iliac crest	Humerus (inner edge of bicipital groove)	Adduction of the shoulder, drawing the arm backwards and downwards, as in bell-pulling and rowing, and internal rotation of the shoulder
Serratus anterior	Over the sides of the thorax but under the scapula in the back	The eight upper ribs in the front of the thorax	The medial border of the scapula	Draws the scapula forwards, antagonistic to the trapezius

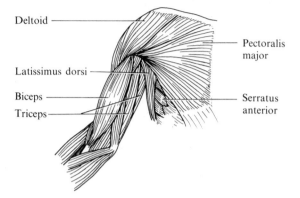

Deltoid

Pectoralis major

Latissimus dorsi

Biceps

Serratus anterior

Triceps

FIG. 81. The muscles of the shoulder and arm.

Muscles of Respiration

The chief muscles of respiration are:

(1) The diaphragm
(2) The external intercostals
(3) The internal intercostals

The *diaphragm* is a dome-shaped sheet of muscle dividing the thorax from the abdomen. The border is of muscle, while the centre is a sheet of fibrous tissue or aponeurosis. The muscle

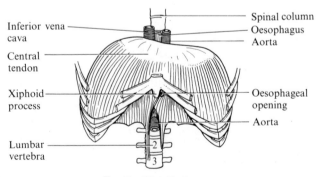

Inferior vena cava

Spinal column

Oesophagus

Aorta

Central tendon

Xiphoid process

Oesophageal opening

Aorta

Lumbar vertebra

FIG. 82. The diaphragm.

arises from the tip of the sternum, the lower ribs and their cartilages, and the first three lumbar vertebrae, and is inserted into the central aponeurosis. Three openings in the diaphragm allow for the passage of the oesophagus, aorta and inferior vena cava, together with lesser structures such as the vagus nerve and thoracic duct, which accompany the oesophagus and aorta respectively. When the muscle fibres contract the dome of the diaphragm is flattened, and lowered, which increases the depth of the thorax from top to bottom.

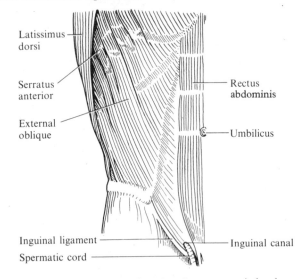

FIG. 83. The abdominal wall. Note that the aponeurosis has been cut away to expose the rectus.

The *external intercostal* muscles lie between the ribs, their fibres running downwards and forwards from one rib to the rib below. They originate from the lower border of the rib above and are inserted into the upper border of the rib below. Their action is to draw the ribs upwards and outwards, to increase the size of the thorax from side to side and from back to front.

The *internal intercostal* muscles also lie between the ribs under the external intercostals and are antagonistic to them.

Their fibres run downwards and backwards from one rib to the rib below. They originate from the lower border of the rib above and are inserted into the upper border of the rib below. Their action is to draw the ribs downwards and inwards to decrease the size of the thorax from side to side and from back to front, particularly in forced expiration.

Muscles Forming the Abdominal Wall

The chief muscles of the abdominal wall are:

(1) The rectus abdominis, forming the front wall
(2) The external oblique ⎫ Forming the side
(3) The internal oblique ⎬ wall and lying one
(4) The transversus abdominis ⎭ under the other
(5) The quadratus lumborum

The *rectus abdominis* forms the front abdominal wall, running up from the pubis to the sternum and costal cartilages. Its fibres run straight up and down, hence its name, for rectus means straight. It is divided into two parts by a line of fibrous tissue in the middle line of the body. This is called the *linea alba*. It is also crossed by lines of fibrous tissue at intervals. These fibrous bands strengthen the muscle and help to prevent stretching.

The *external oblique* muscle forms the outer coat of the side wall. Its fibres run downwards and forwards (cf. the external intercostals). It arises from the lower ribs and is inserted into the iliac crest and the inguinal ligament. The *inguinal ligament* forms the firm edge of the abdominal wall across the groin, where the muscles are not inserted into the bone, leaving a gap for muscles, blood vessels and nerves to pass under it into the limb from the trunk. It gives attachment to the muscles here. It is a strong cord of fibrous tissue. Across the front of the abdomen the external oblique forms a strong aponeurosis which passes in front of the rectus, joining in the linea alba.

The *internal oblique* forms the second coat of the side wall of the abdomen. Its fibres run upwards and forwards. It arises from the iliac crest and the inguinal ligament, and is inserted into the lower ribs and their cartilages. It also forms an aponeurosis, passing partly in front and partly behind the rectus, and joining those of the external oblique and transversus abdominis.

The *transversus abdominis* forms the inner coat of the side wall of the abdomen, lying under the internal oblique. Its fibres run straight round the abdominal wall. It arises from the iliac crest and the lumbar fascia, by which it is joined to the lumbar vertebrae. It is inserted into an aponeurosis which runs across the front of the abdomen, behind the rectus and joins the linea alba.

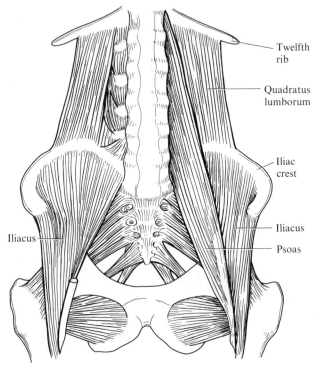

FIG. 84. The iliopsoas.

The *quadratus lumborum* forms the back wall, running up from the iliac crest to the twelfth rib and upper lumbar vertebrae. It holds the twelfth rib steady during breathing.

The abdominal wall is pierced by a canal in either groin; this is called the *inguinal canal*. It runs obliquely through the muscular coats just above the inguinal ligament near its inner extremity. The canal gives passage to structures: in the male the spermatic cord from the testicle, in the female the round ligament of the uterus runs through it with its associated blood vessels and nerves.

Muscles moving the Hip

The muscles in the trunk moving the hip are: (a) The iliopsoas, consisting of (i) the psoas, (ii) the iliacus. (b) The gluteal muscles: major, medius, minimus.

The *iliopsoas muscles* cross the front of the groin, under the inguinal ligament. The psoas arises from the bodies of the lumbar vertebrae, and the iliacus from the front surface of the upper part of the ilium. They are both inserted into the lesser trochanter of the femur. They flex the hip joint and also produce abduction and lateral rotation, but, when the femur is fixed, bend the trunk forwards.

The *gluteal muscles* form the buttocks, running from the back of the sacrum and ilium to their insertion in the greater trochanter of the femur and the gluteal ridge below it. They are three in number—the gluteus maximus, medius and minimus. They extend the hip joint and also abduct and laterally rotate the hip (see Table 5, p. 124), but when the femur is fixed, they extend the trunk on the lower limb. These muscles are commonly used as the site for intra-muscular injections as they are thick, fleshy muscles. Care must be taken to use the upper outer quadrant as the sciatic nerve passes through the other quadrants.

Muscles Moving the Spine

The muscles of the abdominal wall flex and turn the trunk, the rectus flexing and the side muscles turning the thorax on the abdomen. The abdominal muscles also compress the internal organs. The *spinalis* extends the spine. It runs up the back of the trunk on either side of the spine. It arises from the back of the iliac crest and the sacrum, and is inserted into the ribs and upper vertebrae.

Muscles of the Pelvic Diaphragm

The pelvic diaphragm consists of muscles which form the support of the pelvic organs; it runs from the pubis in front, back to the sacrum and coccyx, and out to the ischium on either side. It is in shape like an open book, sloping downwards from back to front and from either side towards the middle line. It is composed of the levator ani muscle and coccygeus muscles. It is pierced in the mid-line in the female by three openings for the passage of the urethra, the vagina, and the rectum. In the

Table 5
Trunk Muscles moving the Hip

Name	Position	Origin	Insertion	Action
Psoas major	Crosses the front of the groin under the inguinal ligament	The bodies of the lumbar vertebrae	The lesser trochanter of the femur	Flexes the hip, abducts and laterally rotates it
Iliacus	Crosses the groin under the inguinal ligament with the psoas	The front surface of the iliac bone	The lesser trochanter	Same as the psoas
Gluteal muscles	Cross the back of the hip, forming the buttocks	The back surface of the ilium and sacrum	The greater trochanter and gluteal line of femur	Extend, abduct and laterally rotate the hip

male there are only two openings, for the urethra and the rectum, respectively.

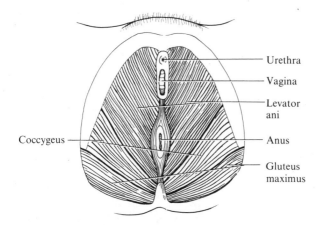

FIG. 85. The pelvic diaphragm.

The Muscles of the Upper Limb

The chief muscles of the upper limb may be divided into: (a) The muscles of the *arm*; and (b) those of the *forearm*; and those of the *hand*.

The Muscles of the Arm

The muscles of the arm are the largest and strongest of the limb and include:

(1) The biceps
(2) The triceps
(3) The deltoid
(4) The brachialis

The **biceps** is so called because it has two heads (Latin *caput* = head). It runs down the front of the arm, where it can easily be felt when it is contracted. It arises by two heads, one from the glenoid cavity and one from the coracoid process of the scapula, and is inserted into the radial tuberosity in the forearm, crossing the front of the elbow joint. It flexes the elbow and also the shoulder, and supinates the hand. Hence

Table 6
Muscles of the Arm

Name	Position	Origin	Insertion	Action
Biceps (with two heads)	In the front of the arm	By two heads from the scapula (coracoid process and above the glenoid cavity)	The radial tuberosity	Flexion of elbow and shoulder and supination of the hand
Triceps (with three heads)	At the back of the arm	By one head from the scapula (axillary border) and two heads from the shaft of the humerus	The olecranon of the ulna	Extension of elbow and shoulder
Deltoid (triangular)	Over the shoulder like an epaulette	The clavicle, acromion process, and spine of the scapula	The deltoid tuberosity of the humerus	Abduction of the shoulder to a right angle
Brachialis	Crossing the front of the elbow	The humerus	The coronoid process of the ulna	Flexion of the elbow

the act of supination can be carried out with great force, and screws and nuts are made so that a right-handed man tightens them with the action of supination.

The **triceps** is so called because it has three heads. It lies at the back of the arm: it arises by three heads, one from the scapula and two from the humerus, and is inserted into the olecranon of the ulna, at the back of the elbow joint. It extends both elbow and shoulder and is antagonistic to the biceps.

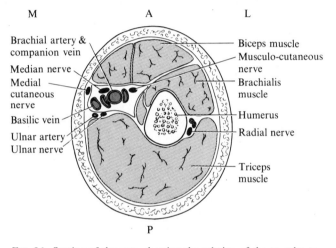

M   A   L

Brachial artery & companion vein

Median nerve

Medial cutaneous nerve

Basilic vein

Ulnar artery
Ulnar nerve

Biceps muscle
Musculo-cutaneous nerve
Brachialis muscle
Humerus
Radial nerve

Triceps muscle

P

FIG. 86. Section of the arm, showing the relation of the muscles to the bone and other structures. L, lateral; M, median; A, anterior; P, posterior aspect.

The **deltoid** is triangular in shape. It lies over the shoulder in the position of an epaulette with the base of the triangle forming the origin and attached to the shoulder girdle, just above the shoulder joint. It is inserted into the deltoid tuberosity on the outer side of the humerus. Its action is to abduct the shoulder to a right angle. (To raise the arm above a right angle, the shoulder girdle must move too, and the trapezius comes into play, drawing scapula and clavicle up towards the occiput.) The front part of the deltoid muscle, used alone, helps to flex the shoulder, moving the humerus forwards, and the back portion, used alone, helps to extend the shoulder, moving the humerus backwards and towards the middle line.

The **brachialis** lies lower in the front of the arm than the biceps, arising from the humerus and being inserted into the coronoid process of the ulna. It assists the biceps in the powerful action of flexion of the elbow (see Table 6).

The Muscles of the Forearm

The forearm contains numerous small, less powerful muscles for movements of the wrist and digits. In the *front* lie the *flexors* of the wrist, the common *flexors* of the fingers, the long *flexor* of the thumb and the *pronator* muscles of the wrist. The com-

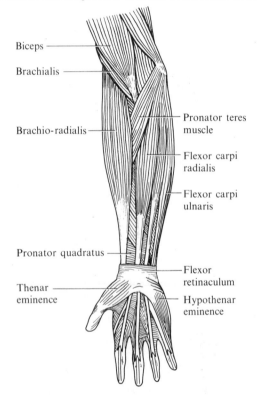

Biceps

Brachialis

Brachio-radialis

Pronator teres muscle

Flexor carpi radialis

Flexor carpi ulnaris

Pronator quadratus

Flexor retinaculum

Thenar eminence

Hypothenar eminence

FIG. 87. The muscles of the forearm and hand.

mon flexors of the digits divide into four tendons, which run across the palm of the hand and up to the terminal phalanx of each digit, into which they are inserted. At the *back* lie the *extensors* of the wrist, the common *extensors* of the fingers, the *extensor* of the thumb and of the first finger, and the *supinators* of the wrist. The tendons of the muscles which cross the wrist are bound down by the flexor retinaculum just above the wrist joint. In the same way the tendons are bound down to the fingers to keep them close to the bones.

The Muscles of the Hand

The hand contains very little muscle, as this would tend to make it clumsy and interfere with its usefulness in grasping and lifting. Many of the muscles which move it are therefore in the forearm. The hand only contains the short flexor of the thumb, and the adductor and abductor muscles for the digits. These latter are termed the interosseous muscles, and are only well developed at the base of the thumb and to a lesser extent at the base of the little finger. Here they form the *thenar* and *hypothenar* eminences, and give power to the grip, where adduction of the thumb is particularly important.

The Muscles of the Lower Limb

The muscles of the lower limb are much larger and more powerful than those of the upper, as the limb carries the whole weight of the body. There is ten times the power in them. They may be divided into: (a) the muscles of the *thigh*, (b) the muscles of the *leg*, and (c) the muscles of the *foot*.

The Muscles of the Thigh

The muscles of the thigh are particularly strong, and include: (a) The *quadriceps femoris*, (b) the *hamstrings*, (c) the *sartorius*, and (d) the *adductors* of the hip.

The *quadriceps* is so called because it has four heads; rather it is four muscles with a common insertion into the patella, and through the patellar ligament it joins the tibia and is the extensor of the knee joint, used in standing and in the powerful action of kicking. It is made up of the *rectus* or straight muscle, and the three *vastus* muscles, the lateral, intermedial and medial. Of these the lateral is the longest muscle and lies on the outside of the thigh. It is occasionally used for the giving of intramuscular injections, as it is well away from the blood vessels, nerves and lymphatics of the limb here.

The *hamstrings* are the flexor muscles of the knee and are so called because of the strong tendons or 'strings' that they form on either side of the popliteal space, at the back of the knee joint. These can be seen and felt readily when the knee is bent.

The muscles are:

(1) The *biceps femoris*, so called because it arises by two heads, one from the ischial tuberosity, the other from the back of the femur. It is inserted into the fibula. It lies on the outside of the back of the thigh.

(2) The *semitendinosus* is so called because of the length of the tendon by which it is inserted into the tibia. It arises from the ischial tuberosity, with the biceps muscle, and lies in the middle of the back of the thigh.

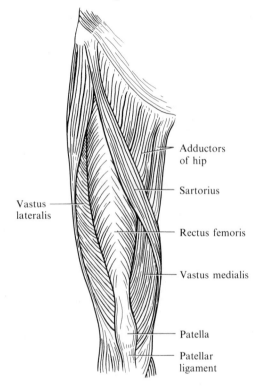

Vastus lateralis

Adductors of hip

Sartorius

Rectus femoris

Vastus medialis

Patella

Patellar ligament

FIG. 88. The muscles of the front of the thigh.

Table 7
Muscles of the Thigh

Name	Position	Origin	Insertion	Action
Quadriceps femoris	Front of thigh	Ilium and femur	Patella, through which it is joined to the tibia by the patellar ligament	Extension of the knee and flexion of the hip
Hamstrings	Back of thigh	Ischial tuberosity and femur	Tibia and fibula by tendons at either side of the popliteal space	Flexion of the knee and extension of the hip
Sartorius	Crosses front of the thigh	Anterior superior iliac spine	Tibia at inner side below the knee	Assists in flexion of hip and knee and abduction and external rotation of hip
Adductors of the hip	Inner side of thigh	Pubis and ischium	Linea aspera of femur and medial epicondyle	Adducts the hip

(3) The *semimembranosus* is so called because of the membranous tendon by which it arises from the ischial tuberosity. It is inserted into the tibia and lies on the inner side of the back of the thigh.

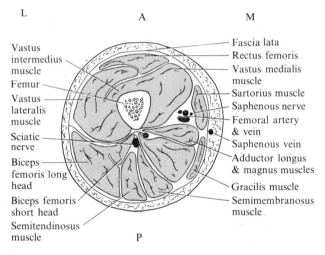

L   A   M

Vastus intermedius muscle —   — Fascia lata
Femur —   — Rectus femoris
  — Vastus medialis muscle
Vastus lateralis muscle —   — Sartorius muscle
  — Saphenous nerve
  — Femoral artery & vein
Sciatic nerve —   — Saphenous vein
Biceps femoris long head —   — Adductor longus & magnus muscles
  — Gracilis muscle
Biceps femoris short head —   — Semimembranosus muscle
Semitendinosus muscle —

P

FIG. 89. Section through the thigh, showing the relation of the muscles to the bone and other structures. L, lateral; M, median; A, anterior; P, posterior aspect.

The biceps thus forms the 'hamstring' tendons on the outer side of the back of the knee and the semitendinosus and the semimembranosus form the 'hamstrings' tendon on the inner side of the back of the knee. The hamstring muscles are again a very powerful group of muscles serving to bend the knee in walking, jumping and climbing, and also, when the tibia is fixed, as in standing, pulling on the ischial tuberosity and helping the gluteal muscles to extend the hip joint.

The *sartorius*, or tailor's muscle, runs from the superior anterior iliac spine across the front of the thigh to the inner side of the knee, which it crosses, and is inserted into the tibia. It assists in the joint movements which occur when sitting tailor-wise, flexing hip and knee and rotating the femur.

The *adductor* muscles form the flesh on the inside of the thigh and are muscles by which the hip is adducted, being assisted by the smaller, more superficial muscle, the gracilis. These muscles are particularly well developed in persons who ride horses, since the rider holds on by the knees, keeping the hips adducted. The adductor muscles run from the pubis and ischium and are inserted into the linea aspera and medial epicondyle of the femur. The adductor magnus forms the greater part of this muscle and a canal runs through it which allows the main artery of the thigh to run from the inner side of the thigh to the back of the knee, where it is well protected (see Table 8).

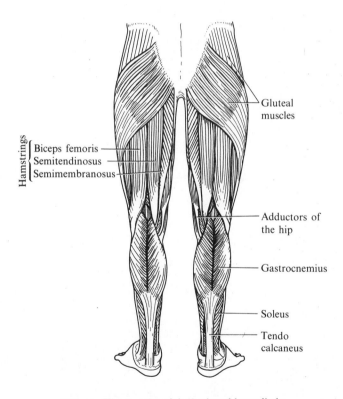

FIG. 90. The muscles of the back and lower limbs.

The Muscles of the Leg

In the leg there are a few large muscles controlling the ankle and many smaller ones moving the foot. The chief muscles are:

(1) The gastrocnemius ⎫
(2) The soleus ⎭ the calf muscles
(3) The tibialis anterior
(4) The flexors and extensors of the digits

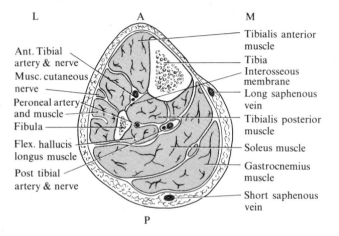

FIG. 91. Section through the leg, showing the position of the muscles in relation to the bones. A, the axis of the centre of the limb; M, median; L, lateral aspect.

The *gastrocnemius* and *soleus* together form the flesh of the calf, the gastrocnemius lying at the back and the soleus lying in front of it. The *gastrocnemius* arises by two heads from the femur, forming the borders of the popliteal space below, as the hamstrings do above the knee. The *soleus* arises from the tibia, not crossing the knee joint and thus not affecting its movements. Both muscles unite below to form a strong common tendon, the tendo calcaneus, by which they are inserted into the calcaneum. The calf muscles raise the heel, causing plantar flexion, or, as it is sometimes called today, extension, of the ankle joint, as in walking and running.

The *tibialis* lies in the front of the leg just outside the crest of the tibia, where it can be seen and felt when the toes and

ball of the foot are raised from the ground. It is this muscle
which gets stiff from the abnormal exercise, when persons
unused to walking up steep slopes begin to climb mountains.
It arises from the tibia and fibula below the knee joint, and is
inserted into the tarsal and metatarsal bones on the inside of
the instep, where its tendon can be seen and felt readily when
the ball of the foot is raised from the ground and the sole turned
to face in towards the middle line. It produces dorsiflexion of
the ankle—i.e. bending towards the dorsal surface—or, as it is
sometimes called today, flexion, bringing the terminology into
line with that used for all other joints. It also inverts the foot
(see Table 8).

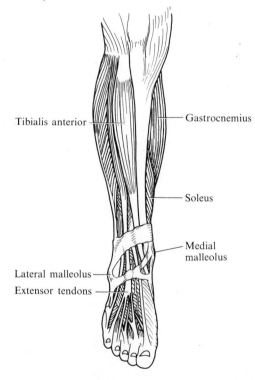

FIG. 92. The muscles of the front of the leg and foot.

Table 8
Muscles of the Leg

Name	Position	Origin	Insertion	Action
Gastrocnemius	Back of the calf	Epicondyles of the femur	The calcaneum by the tendo calcaneus	Plantar flexion of the ankle, raising the heel
Soleus	Front of the calf	Tibia and fibula	The calcaneum by the tendo calcaneus	Plantar flexion of the ankle, raising the heel
Tibialis anterior	Front of the leg	Tibia	Tarsals and metatarsals at the inside of the instep	Dorsiflexion of the ankle, raising the toes, and inversion of the foot

The Muscles of the Foot

The foot, like the hand, contains little muscle, the chief muscles moving the foot lying largely in the leg. The tendons of the extensors of the digits cross the dorsal surface of the foot, the big toe having an individual muscle and tendon. The tendons of the flexors of the digits cross the sole and are strong, and very important in helping to support the arch of the foot. There is a common flexor for the small toes and a flexor for the big toe. In addition, a short flexor of the toes crosses the sole from the calcaneum to the phalanges, and also gives support to the arch. Small interosseous muscles between the metatarsal bones abduct and adduct the digits, but are little used, and therefore little developed.

12 The Blood

THE circulatory system is the transport system of the body by which food, oxygen, water and all other essentials are carried to the tissue cells and their waste products are carried away. It consists of three parts:

(1) The *blood*, which is the fluid in which materials are carried to and from the tissues.

(2) The *heart*, which is the driving force which propels the blood.

(3) The *blood vessels*, the routes by which the blood travels to and through the tissues, and back to the heart.

The blood is a thick, red, fluid; it is bright red in the arteries, where it is oxygenated, and a dark purplish-red in the veins, where it is deoxygenated, having given up some of its oxygen to the tissues—the cause of the colour change—and received waste products taken in from them. It is slightly alkaline in reaction, and the reaction varies very little during life, as the cells of the body can live only if the reaction is normal.

It forms about one-twentieth of the body weight, so that the average volume is 5 to 6 litres.

Composition of the Blood

Although apparently fluid, blood actually consists of a fluid and a solid part. When examined under the microscope, large numbers of small round bodies can be seen in it which are known as the *blood corpuscles* or cells. These form the solid part, while the liquid in which they float is the fluid part and is called *plasma*. The cells form 45 per cent and the plasma 55 per cent of the total volume.

Plasma

The plasma or fluid part of the blood, is a clear, straw-coloured, watery fluid, similar to the fluid found in an ordinary blister. It consists of:

(1) Water, which forms over 90 per cent of the whole.

(2) Mineral salts. These include chlorides, phosphates and carbonates of sodium, potassium and calcium. The chief salt present is sodium chloride, or common salt, of which there are two teaspoonsful (9 grams) to each litre. This is a strength of 0·9 per cent which is called normal saline. The correct balance of the various salts is necessary for the normal functioning of the body tissues.

(3) Plasma proteins; albumin, globulin, fibrinogen, prothrombin and heparin.

(4) Foodstuffs in their simplest forms; glucose, amino acids, fatty acids and glycerol, and vitamins.

(5) Gases in solution; oxygen, carbon dioxide and nitrogen.

(6) Waste products from the tissues; urea, uric acid and creatinine.

(7) Antibodies and antitoxins which protect the body against bacterial infection.

(8) Hormones from the ductless glands.

(9) Enzymes.

The *water* in plasma provides fresh water to supply the fluid which bathes all the body cells and renews the water within the cells. Sixty per cent of the body weight is water and in an average man weighing 70 kg (kilograms) this would be approximately 42 litres. Of the 42 litres two-thirds (28 litres) is within the cells (intracellular fluid) and one-third (14 litres) is outside the cells (extracellular fluid). The extracellular fluid is divided between the blood vessels (6 litres) and the fluid bathing the cells, called the interstitial fluid (6 to 8 litres).

The *salts* in the plasma are necessary for the building of protoplasm and they act as buffer substances neutralizing acids or alkalies in the body and maintaining the correct reaction of the blood. Blood is always slightly alkaline in health and has a pH of 7·4 (see Chapter 1). In plasma there are approximately 155 mEq/l of positively charged ions, chiefly sodium, balanced by 155 mEq/l of negatively charged ions, mainly chloride and carbonate or bicarbonate. This is referred to as the *electrolyte balance*, and is similar in the interstitial fluid. In the intracellular fluid potassium replaces sodium as the negatively charged ions, and phosphate ions and proteins replace chloride as the positively charged ions.

The *proteins* which plasma contains give the blood the sticky consistency, called viscosity, which is necessary to prevent too much fluid passing through the capillary walls into the tissues. If there is a deficiency of protein as in kidney disease when protein is constantly lost as albumin in the urine, the osmotic pressure of the plasma is lowered and excess fluid escapes into the tissues. This excess fluid in the tissues is called *oedema*. The viscosity of the blood also assists in the maintenance of the blood pressure. Albumin is thought to be formed in the liver and globulin is derived from the group of white blood cells called lymphocytes. Fibrinogen and prothrombin are produced in the liver and are both necessary for the mechanism of the clotting of blood. Plasma without fibrinogen is called serum—this can be seen as the yellow fluid which oozes from a cut after a clot has formed. Heparin is also formed in the liver and prevents blood clotting in the vessels.

Foodstuffs, in the form of glucose, amino acids, fatty acids and glycerol, are absorbed from the alimentary tract into the blood. They are the end-products of carbohydrate, protein and fat metabolism.

Urea, uric acid and creatinine are the *waste products* from protein metabolism. They are made in the liver and are carried by the blood for excretion by the kidneys.

Antibodies and Antitoxins are complex protein substances which provide protection against infection and neutralize the poisonous bacterial toxins.

Enzymes are chemical substances produced by the body, which produce chemical changes in other substances without themselves entering into the reaction.

The Blood Cells

The cells are of three types: red blood cells (erythrocytes), white blood cells (leucocytes) and platelets (thrombocytes).

Red Cells

The *red cells* are minute disc-shaped bodies, concave on either side. They are very numerous, numbering about 5 000 000 per cubic millimetre of blood. They are very minute, having a diameter of 7·2 microns only (1 micron = 1 micrometre = 1/1000 millimetre; it is abbreviated to μm or μ). They have no nucleus, but contain a special protein known as *haemoglobin*. This is a pigment and is yellow in colour, though the massed effect of these numerous yellow bodies is to make the blood red. Haemoglobin contains a little iron, and this iron is essential to normal health, though the total amount in the whole body is said to be only sufficient to make a 2-in. nail. Haemoglobin has a great attraction for oxygen. As the red cells pass through the lungs the haemoglobin combines with oxygen from the air (forming oxyhaemoglobin) and becomes bright in colour. This makes the oxygenated blood bright red. As the red cells pass through the tissues oxygen is given off from the blood and the haemoglobin becomes a dull colour (reduced haemoglobin) making the blood a dark purplish-red. The haemoglobin is measured in grammes per 100 ml; the normal figure is 14 to 16 g per 100 ml, or 100 per cent.

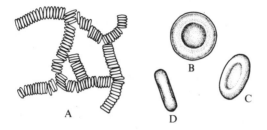

FIG. 93. Red cells. (A) In rouleaux, as seen in shed blood under the microscope; (B), (C) and (D) three views of a single cell.

The *function* of the red cells is therefore to carry oxygen to the tissues from the lungs and to carry away some carbon dioxide. This is their sole function, and is dependent on the amount of haemoglobin they contain. If there is a lack of haemoglobin, either because red cells are reduced in number or because each one does not contain the normal quantity of haemoglobin, the individual suffers from anaemia.

The red cells are *produced* in the red bone marrow of cancellous bone, which is found in the extremities of long bones and in flat and irregular bones. In childhood the red bone marrow also extends throughout the shaft of the long bones, as children have greater need for the production of red cells.

Red cells pass through several stages of development in the bone marrow. Erythroblasts are large cells containing nuclei and a small quantity of haemoglobin. These develop into normoblasts which are smaller cells with more haemoglobin and smaller nuclei. The nucleus then disintegrates and disappears and the cytoplasm contains fine threads, at which stage the cells are called reticulocytes. Finally the threads disappear and the fully mature erythrocyte passes into the blood stream. In health almost all the red cells in the blood should be erythrocytes, with only an occasional reticulocyte. Many factors are necessary for the normal formation of red blood cells.

Protein is required for the manufacture of protoplasm.

Iron is needed for the haemoglobin. Very little iron is excreted. As the red cells are broken down the iron is stored and used again, but a certain amount of iron must be taken in in the diet. A man requires about 10 mg of iron per day, women require about 15 mg to make good the menstrual loss and the depletion of iron reserves which occur during pregnancy, labour and lactation. Iron-containing foods are red meat, eggyolk, green vegetables, peas, beans and lentils.

Vitamin B$_{12}$ (cyanocobalamin) is necessary for the maturation of red blood cells and is usually found in adequate quantities in the diet in temperate climates. It can be absorbed from the small intestine only when it has been combined with the intrinsic factor which is secreted by the stomach. Together these two substances are known as the anti-anaemic factor (or haemopoietic factor), which is stored in the liver and passed to the bone marrow as necessary. Vitamin B$_{12}$ is also known as the extrinsic factor.

Other factors which are necessary even though in small quantities are Vitamin C, folic acid (one of the Vitamin B complex), the hormone thyroxine and traces of copper and manganese.

Red blood cells live in the circulation for about 120 days after which they are ingested by the cells of the reticuloendothelial system in the spleen and lymph nodes. Here the haemoglobin is broken down into its component parts, which are carried to the liver. The globin is returned to the protein stores or is excreted in the urine after further breaking down. The haem is further split into iron, which is stored and used again, and pigment, which is converted by the liver into bile pigments and is excreted in the faeces. Red cell production and breakdown usually proceed at the same rate so the number of cells remains constant.

White Cells

The white cells, or leucocytes, are larger than the red cells, measuring about 10 μm in diameter, and they are less numerous. There are 7000 to 10 000 per cubic millimetre (mm³) of blood, though this number increases considerably, to 30 000/ mm³ when infection is present in the body. This increase is known as leucocytosis. The leucocytes are of three different types.

(1) Polymorphonuclear leucocytes are also known as granulocytes because of the granular appearance of the cytoplasm. The nucleus gradually develops several lobes, hence the name (*poly* = many, *morph* = form). These cells make up about 75

per cent of the total white cells. They are made in the red bone marrow and survive for about 21 days. Granulocytes can be further classified according to their staining properties. Neutrophils (70 per cent) absorb both acid and alkaline dyes. They have the ability to ingest small particles, for example bacteria and cell debris. This power is called phagocytosis and they are sometimes known as phagocytes. They have amoeboid movement and can pass out of the blood stream through the capillary walls to accumulate where there is infection.

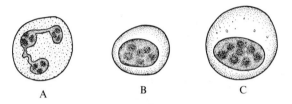

FIG. 94. White cells.
(A) A polymorphonuclear leucocyte; (B) a lymphocyte;
(C) a monocyte.

Eosinophils (4 per cent) absorb acid dyes and stain red. An increase in their number occurs during allergic states such as asthma, and during infestation with worms.

Basophils (1 per cent or less) absorb alkaline or basic dyes and stain blue. They contain heparin and histamine.

(2) Lymphocytes make up about 20 per cent of the total white cell count. They are made in the lymph nodes and in the lymphatic tissue which is present in the spleen, the liver and other organs. They show some amoeboid movement but are not actively phagocytic. They are concerned in the production of antibodies.

(3) Monocytes make up about 5 per cent of the total white cell count. They are the largest of the white blood cells and have a horse-shoe shaped nucleus. They show both amoeboid movement and are phagocytic, and may be members of the reticulo-endothelial system.

Platelets

Platelets, or thrombocytes, are even smaller than red blood cells and are made in the bone marrow. There are about 250 000/mm³ of blood. They are necessary for the clotting of blood.

The Clotting of Blood

When blood flows over a rough surface an enzyme called thrombokinase is released from damaged tissue cells or damaged thrombocytes. In the presence of this enzyme, and

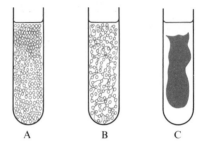

FIG. 95. Clotting of blood.
(A) Blood shed; (B) fibres formed; (C) clot floating in serum.

when adequate amounts of calcium are also present, pro-thrombin, which is a protein normally present in the plasma, is converted into a new substance called thrombin. Thrombin is also an active enzyme which acts on fibrinogen, another of

Table 9

Stages in Clot Formation

(1) Damaged platelets or tissue cells release	thrombokinase
(2) Thrombokinase + prothrombin + calcium	= thrombin
(3) Thrombin + fibrinogen	= fibrin
(4) Fibrin + cells	= clot

the normal plasma proteins, to produce an insoluble thread-like substance called fibrin. The fibrin threads entrap blood cells to form a clot. After some time the clot shrinks and serum is released (serum = plasma − fibrinogen).

Factors Affecting Clotting

Prothrombin is made in the liver and Vitamin K is necessary for its manufacture. Vitamin K is present in green vegetables and is also manufactured in the intestines by bacterial action. It can be absorbed from the intestines into the blood only in the presence of bile. If bile is not present, as in some forms of jaundice, prothrombin may be lacking and the tendency to bleed is increased.

Heparin is a protein normally present in blood, which is formed in the liver and which prevents blood clotting in the vessels. It is called an anti-coagulant.

The Functions of Blood

The uses of blood are:

(1) To carry food to the tissues.

(2) To carry oxygen to the tissues in oxyhaemoglobin.

(3) To carry water to the tissues.

(4) To carry away waste products to the organs which excrete them.

(5) To fight bacterial infection through the white cells and antibodies.

(6) To provide the materials from which glands make their secretions.

(7) To distribute the secretions of ductless glands, and enzymes.

(8) To distribute heat evenly throughout the body, and so regulate body temperature.

(9) To arrest haemorrhage through clotting.

Blood Groups

Blood from one individual cannot always be safely mixed with that of another. This fact became evident with the introduction of blood transfusion, which, at first sometimes cured but sometimes killed the patient. This was due to the fact that blood is of four basic groups. If blood of differing groups is mixed, the red corpuscles may become sticky and form clumps. This is termed agglutination, and is fatal as the clumps of red cells block the blood vessels and obstruct the circulation, and the kidneys are severely damaged by excretion of excessive quantities of pigment from destroyed red cells.

When a person requires a blood transfusion it is necessary

first to find which group he belongs to and then to find a donor belonging to the same group. Blood groups are named according to the presence or absence of substances called *agglutinogens*, which are present in the red blood cells. There are two agglutinogens, called A and B. If A is present the blood group is called Group A, if B is present the blood group is called Group B, if both A and B are present the blood group is called Group AB, and if neither agglutinogen is present the group is called Group O. Having found to which group the patient belongs and found blood donated by a person belonging to the same basic group, a sample of red blood cells from the donor's blood is mixed with some plasma from the patient who is to receive the transfusion (now called the recipient). This is because plasma contains substances called *agglutinins* which cause agglutination of red cells if incompatible blood groups are mixed. Agglutinins are called anti-A and anti-B, and plasma contains all agglutinins which will not affect its own red cells. Therefore the plasma of Group A contains anti-B agglutinin, the plasma of Group B contains anti-A agglutinin, the plasma of Group AB contains no agglutinins, and the plasma of Group O contains both anti-A and anti-B agglutinins. When the donor's red cells are mixed with the recipient's plasma in the laboratory, it can be seen with the help of a microscope whether agglutination occurs. It will be noticed that Group AB has no agglutinins in the plasma and can therefore not cause any red cells to agglutinate; this means that a patient with blood belonging to this group will be able to receive blood from any other group and the group is therefore known as the universal recipient. Group O contains no agglutinogens in the red cells and therefore they cannot be made to agglutinate by the agglutinins in any plasma. Blood belonging to this group can therefore be given safely to a patient belonging to any group and it is known as the universal donor. In practice the compatibility of blood is always checked very carefully before it is given to a patient. This information can be summarized as shown in the table on p. 147.

Rhesus Factor

In addition to ABO grouping there is an additional factor present in the blood of about 85 per cent of the population. It is an agglutinogen called the Rhesus factor. Those who possess the factor are called Rhesus positive (Rh+), the remaining 15 per cent are called Rhesus negative (Rh−). If an Rh− person receives the blood of an Rh+ donor the agglutinogen stimulates the production of anti-Rh agglutinins. If a second Rh+ transfusion were given later the transfused cells would be agglutinated and destroyed (haemolysed) with serious or fatal results to the recipient. This factor can also cause difficulty during pregnancy. If an Rh− mother is carrying an

Blood group	Agglutinogen in red cells	Agglutinin in plasma	Transfusion possible
A	A	Anti-B	Groups A and O
B	B	Anti-A	Groups B and O
AB	A and B	Neither	Any group
O	None	Anti-A and anti-B	Group O only

Rh+ fetus the mother may begin to produce anti-Rh agglutinins which may then destroy the baby's red cells. The baby may overcome this spontaneously, or may require exchange transfusion.

Immunity

When a foreign protein is introduced into the body the immediate response is the production of a substance which will react with it and render it harmless. The foreign protein is called an *antigen* and substances produced in response to the antigen are called *antibodies*. Antigens can be any foreign protein, but common antigens are micro-organisms, some drugs (penicillin is an example), animal and vegetable proteins, including pollens, and foreign tissues such as transplanted organs. The reaction that occurs may be called an antigen–antibody reaction. When it occurs as a response to micro-organisms it is called immunity.

Immunity is a useful defence against infection. Each type of micro-organism which enters the body acts as an antigen and stimulates the production of specific antibody which destroys that antigen and no other. The first time a person comes in contact with the virus which causes measles, antibody is produced which enables the body to overcome the infection and which remains in the blood ready to prevent any further infection by the same type of virus. Immunity may be active, when the cells of the body make the antibody, or it may be passive, when the antibody has been produced in the cells of another person.

Active immunity may be achieved in several ways. *Active*

natural immunity is gained by having the disease, after which the antibody remains in the blood ready to prevent another attack of the same disease. This type of immunity is also developed by what are called sub-clinical infections, in which the body is exposed to small numbers of micro-organisms insufficient in number to give rise to any definite symptoms, but which are sufficient to stimulate antibody production. *Active artificial immunity* is given to children and travellers in order to prevent their getting diseases which would be serious or fatal. An injection of killed micro-organisms, or live ones which have been rendered harmless, is given and the body responds by producing antibodies. In this way active immunity is built up. Harmless toxins are also used to give this type of immunity. Toxins are chemical poisons produced by micro-organisms and when rendered harmless they also act as antigens. Harmless micro-organisms are called vaccines and harmless toxins are called toxoids. Many diseases are prevented by active artificial immunity; some of the most common are whooping cough, diphtheria, measles, smallpox, poliomyelitis and tuberculosis.

	Immunity		
Active		Passive	
Natural	Artificial	Natural	Artificial
(having disease or sub-clinical infection)	(vaccines toxoids)	(from mother)	(antibodies from another human or animal)

Passive natural immunity is gained by the baby before birth as the antibodies are passed from the mother to the fetus. *Passive artificial immunity* is useful in the prevention of disease or in its treatment. The antibodies are produced in another human or in an animal and are injected into the person at risk. Passive immunity is always short-lived as the antibodies are destroyed after a short time.

13 The Heart and Blood Vessels

The Heart

THE heart is a hollow muscular organ which drives the blood through the arteries. It lies in the centre of the thorax between the lungs, but is more to the left of the midline than to the right. It is roughly cone-shaped, its circular base facing upwards and to the right, towards the right shoulder, and its apex pointing downwards, forwards, and to the left, and resting on the diaphragm. The apex lies on the level of the fifth intercostal space, about 8 cm from the midline, a point which the nurse must be able to find to take the apex beat of the heart.

The heart is the size of the closed fist of the individual, measuring about 12 cm in length, 9 cm across and 6 cm in thickness. It weighs about 280 g.

General Structure of the Heart

The heart is divided from base to apex by a muscular septum or partition into two distinct halves, which have no direct communication with each other. Each side is subdivided into two chambers: (a) The atrium, which is the smaller upper chamber of the heart; this is the receiving chamber into which blood comes. (b) The ventricle, which is the larger lower chamber of the heart; this is the discharging chamber by which the blood is driven through the arteries. Each atrium communicates with the ventricle below it on the same side of the heart by an opening known as the atrio-ventricular opening.

Into the right atrium open the superior and inferior venae cavae. From the right ventricle leads the pulmonary trunk. Into the left atrium open the four pulmonary veins. From the left ventricle leads the aorta.

The Function of the Heart

The heart is a *pump* driving the blood through the arteries. It is a double pump, each side being quite distinct from the other. The left side pumps the bright red oxygenated blood, to all parts of the body. Simultaneously the right side pumps the dark purplish de-oxygenated blood to the lungs where it takes in oxygen and gives off carbon dioxide and water.

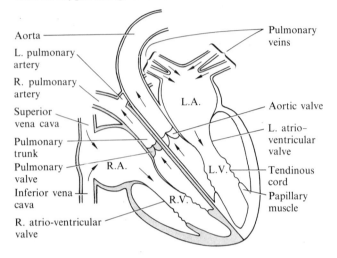

FIG. 96. Diagram showing the structure of the heart. L.A., left atrium; L.V., left ventricle; R.A., right atrium; R.V., right ventricle.

Blood from all parts of the body is collected up into the two large veins, the *superior* and *inferior venae cavae*, which empty into the right atrium. The right atrium contracts when full and drives blood into the right ventricle, which in turn contracts and drives the blood into the *pulmonary trunk*. The pulmonary trunk divides into two branches, the right and the left pulmonary arteries, which carry the blood to the lungs where it takes in oxygen and gives off carbon dioxide and water vapour. The blood is then collected up into four *pulmonary veins*, which empty into the left atrium. The left atrium contracts and drives the blood into the left ventricle. The left ventricle in turn contracts and drives the blood into the *aorta*, which distributes the blood to all parts of the body, where it supplies the tissues with oxygen, water and food and carries away their waste products.

The wall of the heart. The heart wall consists of three coats: (a) the *pericardium*, the outer coat; (b) the *myocardium*, the middle, muscular coat; and (c) the *endocardium*, the inner lining.

The *pericardium* consists of two layers of serous membrane. The inner layer, called the visceral pericardium, covers the heart muscle to the root, where it is turned back to make the outer layer, or parietal pericardium, which forms a loose sac

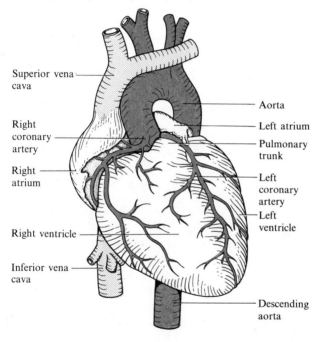

Superior vena cava

Right coronary artery

Right atrium

Right ventricle

Inferior vena cava

Aorta

Left atrium

Pulmonary trunk

Left coronary artery

Left ventricle

Descending aorta

FIG. 97. Front view of the heart.

around the heart. This outer coat contains a good deal of fibrous tissue. The two coats are normally in contact, and their surfaces are smooth and glistening, and are moistened by serum which exudes from the membrane. In normal health there is just sufficient fluid to keep the surfaces moist and prevent friction as the heart beats. In pericarditis, when the pericardium is inflamed there may be an excess of serum, and the amount of fluid in the pericardial sac may seriously embarrass

the heart. The fibrous tissue of the parietal layer of the pericardium prevents over-dilation of the heart, being less elastic
than the muscular coat.

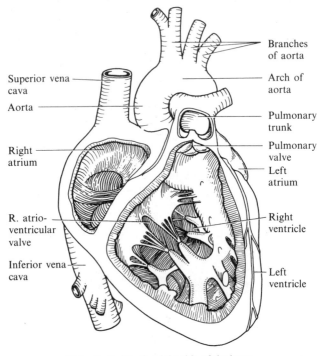

Branches
of aorta

Arch of
aorta

Superior vena
cava

Aorta

Pulmonary
trunk

Pulmonary
valve

Right
atrium

Left
atrium

R. atrio-
ventricular
valve

Right
ventricle

Inferior vena
cava

Left
ventricle

FIG. 98. Inside the right side of the heart.

The *myocardium* is the muscular coat through the action of
which the blood is circulated. It consists of a special type of
muscle, cardiac muscle (see p. 51). This is made up of short
cylindrical fibres, irregularly striped. This muscle is automatic in action, contracting and relaxing rhythmically throughout life without nerve stimulus. In this it differs from the
voluntary muscles of the body. The myocardium varies in
thickness, being thickest in the left ventricle, which has most
work to do, thinner in the right ventricle, which merely drives
the blood to the lungs, and thinnest in the atria.

The *endocardium* is the lining of the heart and consists of a single layer of endothelium, providing a smooth lining for the blood to flow over. The endocardium is continuous with the valves of the heart and with the lining of the blood vessels.

The heart valves. The heart is provided with valves to stop the blood moving in the wrong direction. There are four main valves: (a) the left atrio-ventricular valve between the left atrium and the left ventricle; (b) the right atrio-ventricular valve between the right atrium and the right ventricle; (c) The *aortic* valve between the left ventricle and the aorta; and (d) the *pulmonary* valve between the right ventricle and the pulmonary trunk.

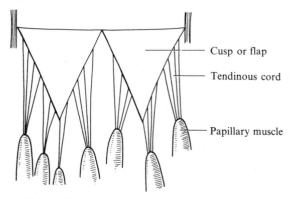

Cusp or flap

Tendinous cord

Papillary muscle

FIG. 99. The mitral valve laid open (diagrammatic).

The left atrio-ventricular valve consists of two triangular flaps or cusps which hang down into the ventricle and are attached by fine tendinous cords (*chordae tendinae*) to muscular projections from the ventricle wall called the papillary muscles. When the ventricle contracts the blood is pushed back towards the atrium but is prevented from entering it by the cusps of the atrio-ventricular valve which is pushed up to block the opening; at the same time the papillary muscles contract and pull on the tendinous cords, which prevent the flaps from being carried too far.

The right *atrio-ventricular valve* is similar in structure and action to the left atrio-ventricular valve, but consists of three flaps instead of two. It prevents the backward flow of blood from right ventricle to right atrium.

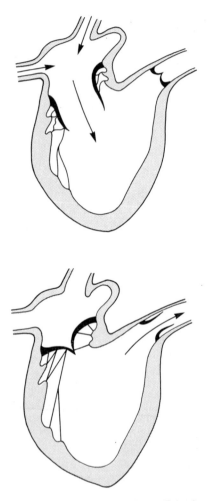

FIG. 100. The action of the valves of the heart. Top: the position of the valves during contraction of the atrium and relaxation of the ventricle. Bottom: the position of the valves during contraction of the ventricle and relaxation of the atrium.

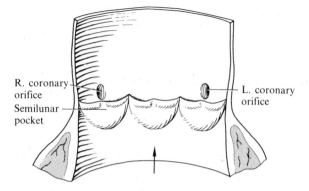

R. coronary orifice

L. coronary orifice

Semilunar pocket

Fig. 101. The aortic valve.

The *aortic valve* consists of three half-moon-shaped cusps at the beginning of the aorta. The cusps are fixed by their curved borders to the vessel wall, the straight edge being free, so that three pockets are formed facing into the aorta. As the blood flows in the right direction from the left ventricle into

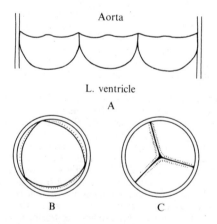

Aorta

L. ventricle

A

B   C

Fig. 102. The aortic valve (diagrammatic).
(a) Laid open, (b) seen from above, and (c) closed.

the aorta the pockets are pushed back flat against the vessel wall; when it tries to return in the wrong direction as the ventricle relaxes the pockets fill with blood, and bulge out, meeting in the centre and completely blocking the opening.

The *pulmonary valve* is similar in structure and action to the aortic, but guards the mouth of the pulmonary trunk preventing the return of blood from it into the right ventricle.

There is also a coronary valve guarding the opening of the coronary vein into the right atrium, and an imperfect valve where the inferior vena cava empties into the right atrium.

Blood supply. The heart is supplied with blood by the *coronary arteries* which branch from the aorta at its commencement and run back into the heart wall. These form a network of capillaries in the wall of the heart. The blood is collected up from these capillaries into one vein, the coronary sinus which empties directly into the right atrium by a special opening guarded by the coronary valve.

The heart beat and nerve supply. The heart beats on an average seventy-two to eighty times per minute in the normal adult, but its rate is increased by exercise and by emotion and excitement. Each beat is a cycle of events. In the first place both atria contract simultaneously, emptying their contents into the ventricles; next, both ventricles contract simultaneously, emptying their contents into the arteries, while the atria begin to relax; the ventricles then also relax and the whole heart is at rest. This rest period lasts as long as the period occupied by the contractions of the atria and ventricles. As a result, in each beat half the period is occupied by the work of contraction of the four chambers and half by rest. The contraction of a chamber is termed *systole*; relaxation is called *diastole*.

When the heart beats rapidly the period of rest is shortened and continued rapid action therefore quickly exhausts the heart. The rest period is also the filling period of the whole heart. When the rest period is shortened over a long period, filling becomes inadequate and circulation fails.

Each beat starts at one point in the right atrium called the 'pace-maker' or sinu-atrial node. This is close to the point where the superior vena cava enters the heart and is made of specialized neuro-muscular tissue. From this point the contraction spreads like a wave over the atria, through similar tissue, to a node between atrium and ventricle, called the atrioventricular node, and then passes from the atria to the ventricles by a special bundle of this tissue in the septum, called the atrio-ventricular bundle. It then spreads similarly over the

ventricular walls, causing them to contract. If the atrio-
ventricular bundle becomes diseased, so that the waves of
contraction cannot pass down it, the ventricles begin to contract independently, but contract very much more slowly than
they do when stimulated from the atria, so that the pulse is
forty or less per minute. This condition is called heart block
and is very serious, as the tissues do not receive a sufficient
blood supply. In other cases some of the stimuli pass down the
bundle, but some do not, so that the beat is irregular; this is
called partial heart block and occurs typically in digitalis
poisoning.

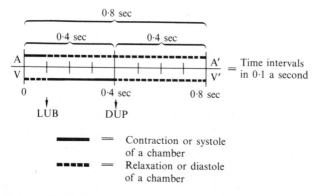

FIG. 103. Graph to show the sequence of events in the cardiac cycle.
A—A¹: the contraction and relaxation of the atrium; V—V¹: the
contraction and relaxation of the ventricle. The cycle occupies 0·8 sec
of which 0·4 sec are occupied by the contraction of both atria and
ventricles, and the remaining 0·4 sec by relaxation of all chambers
of the heart. (Note: 0·8 sec is the time of one beat if the pulse rate is
75 per minute.)

The heart beat is *automatic*, the heart muscle having the
power to contract and relax rhythmically as long as blood circulates through it. The heart is, however, supplied with nerves
which control the rate of the beat, though they do not cause it.
There are two sets of nerves:

(1) The *vagus* nerves, which slow the rate and lessen the
strength of the beat, acting like a brake on the heart.

(2) The *sympathetic* nerves, which quicken and increase the
strength of the beat, acting like the accelerator in a motor car.

The heart beat produces sounds which can be heard by applying the ear or a stethoscope to the chest wall over the heart. The normal sounds are likened to lūb, dŭp, followed by a pause. They are made by the closing of the valves in each heart cycle. Disease of the valves and dilatation of the heart produce abnormal sounds called murmurs.

In health the rate of the heart beat varies:

(1) Rest slows the rate and exercise quickens it.

(2) Age. It is quicker in infants (120 to 130 at birth), and slows as the child grows, and is slow in old age.

(3) Sex. It is quicker in the female sex (80) and slower in the male (72).

(4) Size. It is quicker in smaller animals, slower in large.

(5) Emotions and excitement quicken the pulse.

(6) Temperament affects the pulse. The placid person has a slow pulse, which does not readily alter; the excitable, a quicker pulse which fluctuates readily.

In disease many conditions, such as fever, haemorrhage, shock and hyper-thyroidism, quicken the heart beat, while pressure on the brain, jaundice and heart block slow it. Drugs can also both quicken and slow the beat.

The Blood Vessels

The blood vessels are hollow tubes which carry the blood through the tissues. In health blood is always contained in blood vessels and does not pass out into the tissues except when there is injury or disease.

There are *three* types of blood vessel: (a) arteries, (b) veins, and (c) capillaries.

The *arteries* are stout-walled vessels which carry the blood away from the heart. They carry bright red oxygenated blood except in the case of the pulmonary arteries, which carry de-oxygenated blood. One large artery, the aorta leaves the left ventricle and gives off smaller branch arteries to the various parts; these branch again and again to form finally very fine arteries known as arterioles. These very fine arterioles divide into capillaries.

The *veins* are thin-walled, collapsible vessels which carry blood back to the heart from the tissues. They carry the dark purplish deoxygenated blood except in the case of the pulmonary veins, which carry oxygenated blood. The capillaries unite to form small veins or venules. These unite with other small veins to form larger veins which ultimately unite to form two large veins, the superior and inferior venae cavae, which empty into the right atrium of the heart.

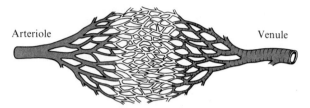

Network of capillaries

FIG. 104. Diagram of a capillary network.

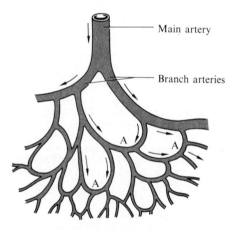

FIG. 105. Diagram to illustrate the anastomosis of arteries.
(A) Anastomosing vessel.

The *capillaries* are fine, hair-like vessels, microscopic in size. They join the arterioles to the venules. They receive blood from the arterioles and pass it into the venules. Their walls are one cell in thickness and through them food, oxygen and water pass out from the blood to the tissue cells and waste products such as carbon dioxide and water can pass back from the tissues into the blood stream. In this way they are unlike either of the other vessels. Arteries and veins merely carry blood to and from the tissues. The blood within them cannot even nourish their own walls, which are themselves supplied with capillaries. The capillaries nourish the tissues. Every tissue has its network of capillaries, except in one or two cases where the tissue is nourished by the blood supply of neighbouring tissues; for example, the outer layer of skin, called the epidermis.

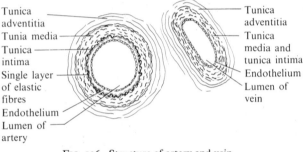

FIG. 106. Structure of artery and vein.

Arteries branch like trees, while veins join together like streams forming a river; but the analogy is not quite correct. The branching arteries link up with one another, and this joining up of the branching arteries is known as *anastomosis*. This is valuable in cases of injury. If the artery supplying one part is cut, blood can pass to it by means of the linking-up vessels; otherwise the part would be completely deprived of blood and would die (see Fig. 105).

Structure of the Blood Vessels

The **arteries** have stout walls which are necessary to withstand the pressure of the blood within them. The walls consist of three coats:

(1) An outer sheath of fibrous tissue, called the tunica adventitia.

(2) A thick coat composed of (a) involuntary muscle, and (b) elastic fibrous tissue, called the tunica media.

(3) A lining of endothelium, which provides a smooth surface for the blood to flow over, called the tunica intima.

There are two distinct types of artery: (a) the *large* artery, which distributes blood to the various organs; and (b) the *arteriole*, which divides into capillaries within the various organs.

The *large elastic artery* has in its thick middle coat a great deal of elastic fibrous tissue and little involuntary muscle. As a result its wall is essentially elastic in character. The aorta and all its main branches are of this type, and because of their elasticity *pulsate* with the heart beat. When the left ventricle contracts, it drives blood into the aorta, which is already full and becomes distended to accommodate the extra blood. When the left ventricle relaxes and dilates and the aortic valve closes the elastic aorta recoils and returns to its original size. The recoil of the aorta is very important in physiology. The heart, sends out an intermittent stream of blood, discharging blood only when it contracts. The tissues need a continuous supply, and this is provided because the recoiling aorta drives blood on while the heart is filling, though not with such force as during the contraction of the ventricle. The stream of blood in the large arteries is therefore irregular, the blood flowing more quickly when the ventricle is contracting and more slowly when the aorta is recoiling.

The Pulse

The distension and recoil of the aorta sets up a wave of distension and recoil which travels along all the large elastic arteries and causes the pulse. The *pulse* is merely a wave of distension, not a wave of blood travelling along the arteries. All the large elastic arteries pulsate, though pulsation can only be felt where a large artery passes over a bone close under the skin; an example is the radial artery which passes over the radius at the wrist. Since the heart beat produces the pulsation, the rate and character of the beat can be judged by taking the pulse. The pulse will vary with the heart beat (see p. 158).

The arteriole has in its thick middle coat a great deal of involuntary muscle and little elastic fibrous tissue. It is therefore essentially muscular in character and the muscle is provided with nerves, called the *vasomotor* nerves, through which the arteries can be *dilated* or *contracted* according to the needs of the tissues. The muscle in the walls of the arterioles is also affected by hormones in the blood, particularly by adrenaline and noradrenaline from the adrenal glands, which cause contraction of the small arteries. These arterioles do not pulsate. When an organ is at work it needs a larger blood supply, and the arterioles dilate so that they carry much blood to the capillaries which nourish it. When the organ is at rest it needs less blood, and the arterioles contract so that they carry less blood to its capillaries. This can be seen in the case of the skin. When the skin is active, getting rid of heat, as in exercise, the blood vessels are dilated and the skin can be seen to be flushed with blood. When the body is at rest and less heat is to be lost, the blood vessels contract and the skin loses its red flush.

The arterioles therefore control the distribution of blood to the various organs of the body. The small blood vessels are normally contracted throughout the body, except in any organ which is active. This contraction of the arterioles is important in maintaining the blood pressure, as the contraction offers resistance to the outflow of the blood from the large elastic arteries.

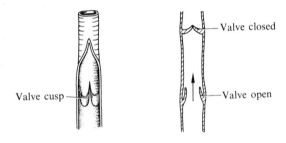

FIG. 107. Valves in the veins.

The **veins** have thinner walls than the arteries, but have the same three coats. The middle coat is comparatively thin, and contains less muscle and elastic tissue so that the veins easily collapse. Many of the veins, especially in the limbs, are provided with *valves* to stop the blood flowing in the wrong direction. Valves are unnecessary in the arteries, as the pressure is so great that the blood can flow only one way, but in the veins the pressure is very slight and valves are necessary, especially in veins which run uphill throughout their course in the standing position. The valves consist of two pockets similar to those of the aortic valve. If a vein is very distended the pockets do not meet across it, and blood can flow either way. This is known as a *varicose* vein.

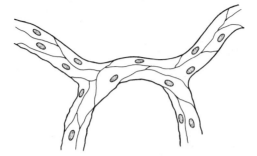

FIG. 108. Capillary network showing the wall of simple endothelium.

The *capillaries* have very thin walls consisting of a single layer of cells only, continuous with the lining endothelium of the arterioles and venules with which they join. Through this single layer of flat cells food, oxygen and water can pass out to the tissue cells; white cells can also push their way between the cells by amoeboid movement, and pass out into the tissues to fight bacterial infection.

14 The Circulation

BLOOD is in constant circulation within the blood vessels throughout the body. In health it does not leave the blood vessels unless they are injured. Oxygenated bright red blood is driven by the left side of the heart through arteries to the tissues. As it passes through the capillaries in the tissues it gives off food and oxygen, and takes in waste products, becoming a dark purplish-red, because it has given up some of its oxygen. This blood is carried back by the veins to the right side of the heart, which drives it to the lungs to be re-oxygenated; from the lungs it is returned to the left side of the heart for redistribution to the body.

The circulation is divided into four parts:

(1) The *general* or systemic circulation
(2) The *pulmonary* circulation
(3) The *portal* circulation ⎫ Branches of the general
(4) The *coronary* circulation ⎭ circulation

The *general circulation* is the distribution of bright red blood rich in oxygen to the tissues and the return of dark purplish blood, depleted of oxygen, to the heart. The aorta leaves the left ventricle and gives off branches to all the organs of the body, from which the blood is returned by veins. These ultimately empty into the superior or inferior vena cava, which return the blood to the right atrium.

The *pulmonary circulation* is the distribution of the dark purplish de-oxygenated blood to the lungs from the right side of the heart, and the return of the bright red oxygenated blood to the left side of the heart. The blood is carried by the right and left pulmonary arteries into the lungs. They divide within the lungs into smaller and smaller arteries which ultimately divide into capillaries. As the blood passes through the capillaries it takes in oxygen and gives off carbon dioxide—and is then gathered up into the four pulmonary veins which leave the lungs and empty the blood into the left atrium.

The *portal circulation* is a branch of the general circulation dealing with blood rich in food-stuffs from the abdominal organs. Blood from the stomach, intestines, spleen and pancreas is gathered up into one large vein, the portal vein, which runs into the liver and there divides into a second set of capillaries. As the blood passes through these capillaries excess food not required for the immediate needs of the body is removed and stored in the liver cells until required for use or is altered as required for body metabolism. The blood passes from the capillaries into veins, which unite to form three large hepatic veins and a number of smaller ones, which empty into the inferior vena cava, carrying into the circulation the food required for immediate use. The hepatic veins also bring back the blood brought by the hepatic artery to supply the liver with oxygen.

The *coronary circulation*, which is also a branch of the general circulation, supplies the myocardium. The right and left coronary arteries leave the aorta at its commencement and run back into the myocardium, where they form a network of capillaries to supply the heart with blood. During contraction of the ventricles, when the aortic valve is open, the entrances to the coronary arteries are covered by the cusps of the aortic valve. The coronary arteries can, therefore, only receive blood during diastole. From the capillaries the blood is collected into one vein, the coronary sinus, which empties directly into the right atrium by a special opening which is guarded by a valve, the coronary valve. This valve protects the coronary vessels from being filled by back pressure with de-oxygenated blood when the atrium contracts.

The Chief Blood Vessels of the General Circulation

The Arteries

The names of the arteries are taken from the bones in the limbs or from the organs they supply. The names are not as important as are their positions, especially their positions in relation to the bones for the taking of the pulse, applying pressure to them in first aid in case of haemorrhage, and preventing pressure on them by splints and other equipment. The arteries are found in the safest positions, on the inner side where there is one bone in the limb, and between the bones where there are two. They cross the flexor surfaces of joints where they are not exposed to pressure and injury.

The *aorta* is the main artery from which branches are given off to all parts of the body. It is a stout vessel about 45 cm long and the thickness of an ordinary garden hose-pipe. It arises from the left ventricle of the heart, running upwards for a short distance as the ascending aorta, then arches over the heart from right to left, forming the arch of the aorta. It then runs down the thorax behind the heart as the descending thoracic aorta, pierces the diaphragm, and runs down the back of the abdomen, forming the abdominal aorta. It ends on the level of the iliac crest, where it divides into two branches, the right and left common iliac arteries.

The *ascending aorta* gives off the right and left *coronary arteries*, which supply the heart wall.

The *arch of the aorta* gives off:

(1) The brachiocephalic trunk, which divides into: (a) the *right subclavian* artery supplying the right upper limb, and (b) the *right common carotid* artery, supplying the right side of head and neck.

(2) The *left common carotid artery*, supplying the left side of head and neck.

(3) The *left subclavian* artery, supplying the left upper limb.

The *descending thoracic aorta* gives off intercostal arteries to supply the intercostal muscles, the bronchial arteries to supply the lungs with fresh blood, and the oesophageal artery supplying the oesophagus.

The *abdominal aorta* supplies the abdominal wall and the abdominal organs.

The chief branches are:

(1) The *phrenic* arteries, supplying the diaphragm.

(2) The *coeliac* trunk, which divides into: (a) the *gastric* artery, supplying the stomach; (b) the *hepatic* artery, supplying the liver, and stomach; and (c) the *splenic* artery, supplying the spleen, and pancreas.

(These arteries anastomose freely with one another, and also provide branches for the duodenum and pancreas.)

(3) The *superior mesenteric* artery, supplying the small intestine and the beginning of the large.

(4) The *renal* arteries, supplying the kidneys.

(5) The *ovarian* or *testicular* arteries, supplying the ovaries and testes respectively.

(6) The *inferior mesenteric* arteries, supplying the remainder of the large intestine.

The *common iliac* arteries divide into:

(1) The *external iliac* artery, supplying the lower limb.
(2) The *internal iliac* artery, supplying the pelvic organs and buttocks.

Blood supply to the head and neck. The head and neck are supplied by:

(1) The *common carotid* arteries.
(2) The *vertebral* arteries, branches of the subclavian arteries.

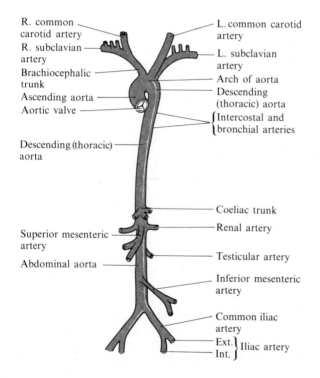

R. common carotid artery
R. subclavian artery
Brachiocephalic trunk
Ascending aorta
Aortic valve
Descending (thoracic) aorta

L. common carotid artery
L. subclavian artery
Arch of aorta
Descending (thoracic) aorta
Intercostal and bronchial arteries

Coeliac trunk
Renal artery
Superior mesenteric artery
Testicular artery
Abdominal aorta
Inferior mesenteric artery
Common iliac artery
Ext. ⎱
Int. ⎰ Iliac artery

FIG. 109. The arch of the aorta and branches.

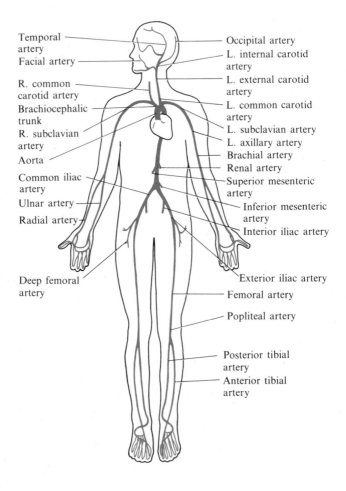

Temporal artery
Facial artery
R. common carotid artery
Brachiocephalic trunk
R. subclavian artery
Aorta
Common iliac artery
Ulnar artery
Radial artery
Deep femoral artery

Occipital artery
L. internal carotid artery
L. external carotid artery
L. common carotid artery
L. subclavian artery
L. axillary artery
Brachial artery
Renal artery
Superior mesenteric artery
Inferior mesenteric artery
Interior iliac artery
Exterior iliac artery
Femoral artery
Popliteal artery
Posterior tibial artery
Anterior tibial artery

FIG. 110. Main arteries.

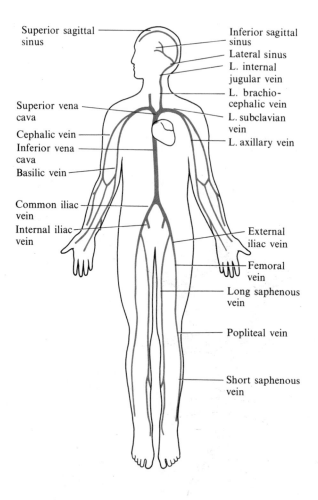

Superior sagittal sinus

Inferior sagittal sinus

Lateral sinus

L. internal jugular vein

L. brachio-cephalic vein

Superior vena cava

L. subclavian vein

Cephalic vein

L. axillary vein

Inferior vena cava

Basilic vein

Common iliac vein

Internal iliac vein

External iliac vein

Femoral vein

Long saphenous vein

Popliteal vein

Short saphenous vein

FIG. 111. Main veins.

The *common carotid* arteries run up the neck on either side of the trachea to the level of the thyroid cartilage, where they divide on either side into:

(1) The *external carotid* artery, which supplies the outer parts of the face and scalp, dividing into many branches such as the facial, temporal and occipital arteries.

(2) The *internal carotid* artery, which supplies the brain and eye.

The *vertebral* arteries run up the neck through the foramina in the transverse processes of the cervical vertebrae (see pp. 77 and 78) and enter the skull by the foramen magnum to supply the brain.

The brain is therefore supplied with blood by four arteries, the right and left internal carotids and the right and left vertebral arteries. These four vessels anastomose freely at the base of the brain, forming the cerebral arterial circle, from which the cerebral arteries—anterior, middle, and posterior—are given off (see Fig. 112). The free anastomosis enables blood to reach all parts of the brain should one of these vessels be injured.

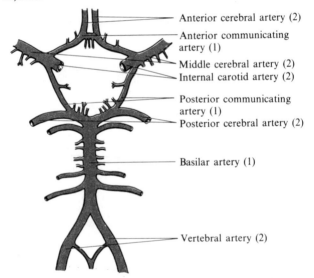

Anterior cerebral artery (2)
Anterior communicating artery (1)
Middle cerebral artery (2)
Internal carotid artery (2)
Posterior communicating artery (1)
Posterior cerebral artery (2)
Basilar artery (1)
Vertebral artery (2)

FIG. 112. The arterial circle of the cerebrum (circle of Willis).

Blood supply to the upper limb. Blood is supplied to the upper limb by the subclavian artery, which runs out over the first rib and under the clavicle to the axilla, where it is called the *axillary* artery. This runs into the arm and becomes the *brachial* artery, which runs down on the inner side of the humerus, passes across the front of the elbow, and divides into the *radial* and *ulnar* arteries in the forearm. The *radial* artery runs down the outside of the limb, crosses the wrist, passing round the back

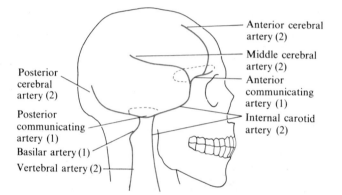

Fig. 113. Blood supply to the brain (side view). Broken lines indicate communication with the opposite side of the brain.

of the base of the thumb to avoid pressure, and enters the palm of the hand to form the deep palmar arch. The *ulnar* artery runs down the inner side of the forearm, crosses the wrist and forms the superficial palmar arch. The *palmar arches* give off small arteries to supply the digits. Each arch anastomoses freely with the other, so that, if one artery be cut, blood can be brought by the other to the parts supplied by the injured vessel and death of tissue is prevented. The disadvantage of this arrangement is that injury to the main artery here causes bleeding from both sides.

Blood supply to the lower limb. Blood is supplied to the lower limb by the *external iliac* artery, which crosses the centre of the groin and runs into the thigh, becoming the femoral artery. The *femoral* artery runs down the inside of the thigh, giving off the deep femoral artery to supply the thigh muscles. The superficial femoral artery runs to the back of the knee, where it is known as the *popliteal* artery. This passes into the leg and divides into:

(1) The anterior tibial artery.
(2) The posterior tibial artery.

The *anterior tibial* artery runs down the front of the leg, passes over the front of the ankle, and forms the dorsalis pedis artery which supplies the dorsum of the foot and lies midway between the malleoli. The *posterior tibial* artery runs down the back of the leg, crosses the ankle on the inner side, half-way between the medial malleolus and the calcaneum, where it is well protected from pressure and injury, and forms an arch to supply the sole of the foot, the plantar arch, formed from the plantar artery and its branches.

The arches in the foot are not so clearly defined as those in the hand, but are similar on general lines and anastomose freely in the same way.

The Veins

There are **two** sets of veins in the general circulation: (a) the deep veins; and (b) the superficial veins.

The *deep* veins accompany the main arteries, and are known as *companion* veins. As a rule they bear the same name as the artery they accompany, but there are exceptions.

The *superficial* veins run in the subcutaneous fatty tissue, and are most numerous in the limbs. They empty finally into the deep veins. They can be seen as blue lines through the skin, the purplish colour of the blood they contain giving this effect when seen through the transparent skin.

The return of blood from the head and upper limbs. The blood from the brain is collected into channels which run over the inside of the cranium, and are known as venous *sinuses*. These sinuses empty into two chief sinuses known as the transverse sinuses, which empty into two large veins in the neck, known as the *internal jugular* veins. These are companion veins to the internal and common carotid arteries. There are also smaller external and anterior jugular veins which help to collect blood from the outer parts of the head, but are comparatively small and unimportant. At the root of the neck the jugular vein joins with the subclavian vein, bringing back blood from the upper limb (Fig. 115).

The blood from the upper limbs is collected by both deep and superficial veins, the superficial ones being the larger.

The blood from the hand is collected into the plexus of veins from which three main superficial veins run up the forearm,

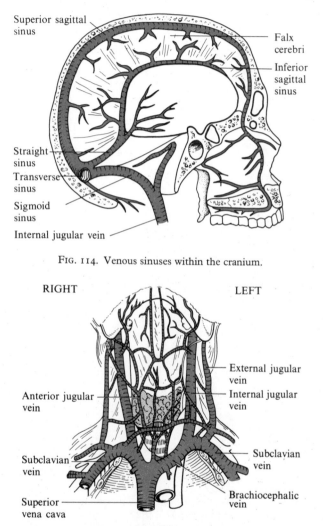

Superior sagittal sinus

Falx cerebri

Inferior sagittal sinus

Straight sinus

Transverse sinus

Sigmoid sinus

Internal jugular vein

FIG. 114. Venous sinuses within the cranium.

RIGHT   LEFT

External jugular vein

Internal jugular vein

Anterior jugular vein

Subclavian vein

Subclavian vein

Superior vena cava

Brachiocephalic vein

FIG. 115. The chief veins of the neck.

called the *cephalic, median* and *basilic* veins. At the elbow the median vein divides into the *median basilic* and *median cephalic* veins. The median basilic joins the basilic vein to form the largest vein in the arm. This runs up on the inner side of the limb and joins the smaller deep brachial veins to form the axillary vein in the axilla. The median cephalic vein joins the *cephalic* vein, runs up the outer side of the arm, crosses the shoulder and finally joins the axillary vein to form the subclavian vein.

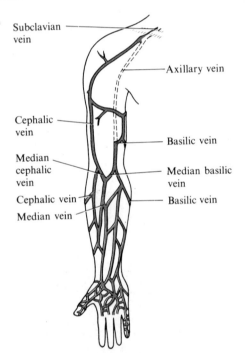

FIG. 116. The superficial veins of the forearm.

The *subclavian* veins, right and left, run from the axillae under the clavicle to the root of the neck, where they join the internal jugular veins on either side to form the right and left brachiocephalic veins. These in turn unite to form the *superior vena cava*, which empties the blood into the right atrium of the heart.

The return of blood from the lower limbs and the abdomen. The blood from the lower limbs is collected by deep and superficial veins. The deep veins are the *anterior* and *posterior tibial* veins, the *popliteal* vein and the *femoral* vein, which accompany the arteries. The main superficial veins are the short and the long saphenous veins. The small or *short saphenous* vein arises on the outer side of the instep and runs up the back of the leg, receiving many tributary vessels, and empties into the deep popliteal vein at the back of the knee.

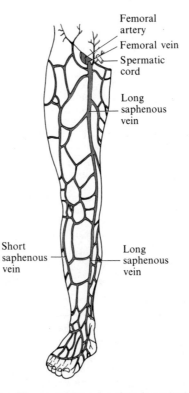

Femoral
artery

Femoral vein

Spermatic
cord

Long
saphenous
vein

Short
saphenous
vein

Long
saphenous
vein

FIG. 117. The superficial veins of the lower limb.

The large or *long saphenous* vein arises on the inside of the instep and runs up the inside of the leg and thigh, receiving many tributary vessels. It empties into the deep femoral vein at the groin just before this passes under the inguinal ligament. It is the longest superficial vein in the body and the one which most often becomes varicose. Between the deep and superficial veins in the leg there are a number of perforating veins, usually three between knee and ankle medially and one near the ankle laterally. These allow the blood from the superficial tissues to drain into the deep veins owing to the strong aspirating action of the calf muscle pump. Valves guarding the entrance of these perforating veins prevent back pressure of blood into the superficial tissues during muscle contraction. If these valves are damaged by venous thrombosis, post-thrombotic varicose ulcers occur.

The *femoral* vein finally crosses the groin alongside the femoral artery, and becomes the *external iliac* vein. This joins the *internal iliac* vein, which returns the blood from the pelvic organs and buttocks, to form the *common iliac* vein on either side. The right and left iliac veins join to form the *inferior vena cava* at the level of the fifth lumbar vertebra where the abdominal aorta ends. The inferior vena cava runs up the back of the abdomen immediately to the right of the abdominal aorta. It receives:

(1) The *renal* veins from the kidneys.

(2) The *ovarian* or testicular veins from the reproductive organs emptying into the renal vein on the left but directly into the vena cava on the right.

(3) The three large *hepatic* veins, which return all the blood from the portal circulation in addition to the blood distributed to the liver.

The inferior vena cava finally pierces the diaphragm and empties into the right atrium of the heart.

The Mechanism of the General Circulation

Blood is driven through the *arteries* by:

(1) The contraction of the left ventricle.

(2) The recoil of the elastic aorta, which drives blood on as the heart fills (see p. 161).

The arterioles are kept in a state of contraction by the adrenaline and noradrenaline in the blood stream. These substances are secreted by the adrenal glands and vary in quantity. Strong emotion stimulates the secretion of adrenaline, and therefore causes a rise in blood pressure through the greater degree of contraction in the arterioles. On the other hand, in Addison's disease the adrenal glands are destroyed and there is a deficiency of adrenaline and noradrenaline in the blood, resulting in a low blood pressure. Similarly, in shock there is an abnormal dilatation of the small blood vessels in the internal organs and therefore a low blood pressure.

Blood is circulated through the *veins* by:

(1) Suction as the heart expands.
(2) Suction towards the thorax on inspiration which sucks blood towards the heart as well as drawing air into the lungs.
(3) Pressure from without on the thin-walled veins due to contraction of the muscles. This massages the veins and drives the blood towards the heart because of the numerous valves found in most of the veins.

Blood Pressure

The *blood pressure* is the force which the blood exerts on the walls of the blood vessels. It varies in the different *blood vessels* and also with the *heart beat*. The pressure is greatest in the large arteries leaving the heart, and gradually falls in the arterioles until, when it reaches the capillaries, it is so slight that the least pressure from without will obliterate these vessels and drive the blood out of them. This can be seen by pressing lightly on the nail or letting a piece of glass rest lightly on the skin. (It is for this reason that it is so important to change frequently the position of a patient confined to bed, as the tissue carrying the body's weight has little blood circulating through it.) In the veins the pressure is lower still until ultimately in the big veins approaching the heart there is suction, i.e. a negative pressure instead of a positive one, on account of the suction exerted by the heart as its chambers dilate.

The pressure in the large arteries varies with the *heart beat*. It is highest when the ventricle contracts (this is called *systolic pressure*) and lowest when the ventricle relaxes (this is called *diastolic pressure*).

The pressure of the blood is measured by the height of the column of mercury which it will support. The height is calculated in millimetres. The normal arterial pressure is *110 to 120 millimetres* systolic pressure and *65 to 75 millimetres* diastolic pressure. The apparatus for measuring blood pressure is called a sphygmomanometer. There are different types, but the most reliable consists of a hollow rubber cuff encased in material which is fixed evenly round the limb without constricting it in any way. To this is attached a small hand pump, by which air can be pumped into the cuff, expanding it and compressing the limb and the blood vessels it contains. The air cushion is also linked to a manometer containing mercury, so that the pressure of the air within it can be read against a graduated scale.

To take the blood pressure the air cuff is fixed round the arm and the pulse at the wrist is felt with one hand while the other is used to inflate the cuff to a pressure above the point at which the radial pulse can no longer be felt. A stethoscope is placed over the brachial artery pulse in the cubital fossa and the pressure in the cuff is slowly lowered by means of the release valve. As the pressure falls no sounds are heard until the systolic blood pressure is reached, at which point a tapping

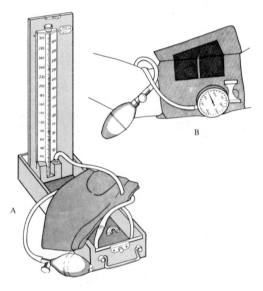

FIG. 118. (A) Diagnostic sphygmomanometer, and
(B) an aneroid sphygmomanometer.

sound is heard in the stethoscope. At this point the level of mercury in the manometer must be noted. As the pressure in the cuff is further reduced the sounds become louder, until the diastolic blood pressure is reached, at which point the sound is altered in character and becomes muffled. A little further decrease in cuff pressure causes the sound to disappear completely. The diastolic blood pressure is noted as the point at which the character of the sound changes.

Arterial Pressure

The arterial pressure is maintained by:

(1) The cardiac output.
(2) The peripheral resistance.
(3) The total blood volume.
(4) The viscosity of the blood.
(5) The elasticity of the arterial walls.

The *cardiac output* is the amount of blood pumped out with each ventricular contraction. When the left ventricle contracts about 70 ml of blood is pushed into the aorta, which is already full, causing it to distend.

The *peripheral resistance* is the resistance offered by the small blood vessels, particularly the arterioles, to the flow of blood. This resistance prevents blood flowing too quickly into the capillaries and in this way helps to maintain the blood pressure in the arteries. The lumen of the arterioles can be altered by the action of vaso-motor nerves and by adrenaline and noradrenaline from the adrenal glands. If the lumen is narrowed the resistance to blood flow is increased and arterial blood pressure is raised. If the lumen is enlarged more blood will pass through quickly to the capillaries and the blood pressure will be lowered.

The *blood volume* is the total amount of circulating blood. If this is reduced by the loss of whole blood, as in haemorrhage, or by loss of fluid from the circulation, as in shock, burns or dehydration, the blood pressure will be lowered.

The *viscosity of the blood* is its stickiness. Blood is a viscous fluid which offers two or three times as much resistance as plain water. The viscosity depends partly on the plasma, particularly the plasma proteins, and partly on the number of red cells. The viscosity of the blood will be reduced if large quantities of intravenous normal saline are given, and lowered viscosity is associated with lowered blood pressure.

The *elasticity of the arterial walls* allows distension of the aorta when the ventricle contracts and elastic recoil as the ventricle relaxes. This recoil pushes the blood on and maintains the diastolic pressure of the blood. Distension and recoil occur throughout the arterial system. Elasticity may be reduced in atheroma which is a degenerative disease of the arteries. In this condition the blood pressure will be raised because of loss of elasticity.

A *high blood pressure* is a systolic pressure of 150 to 180 mm and over. The diastolic pressure is also raised as a general rule, and a high diastolic pressure, for example 90 mm or more, is injurious, as it is a strain on the heart. A high blood pressure is dangerous in that the raised pressure may cause rupture of a blood vessel. Rupture is particularly likely to occur in the brain and is one of the causes of cerebro-vascular accident or stroke. There is also strain on the heart which may result in heart failure.

A *low blood pressure* is a systolic pressure of 100 mm or less. It occurs in cases of haemorrhage, shock and collapse, heart failure, and in disease of the suprarenal glands. The danger lies in an insufficient supply of blood to the vital centres of the brain. The treatment is therefore:

(1) To place the patient flat with the foot of the bed raised if necessary, to send blood by gravity to the vital centres of the brain.

(2) To give heart stimulants—e.g. nikethamide, adrenaline or noradrenaline—and circulatory stimulants such as adrenaline and noradrenaline to contract the blood vessels.

(3) To give fluids such as saline by intravenous, subcutaneous, or rectal infusion, or blood or plasma transfusion, to increase the fluid in circulation.

Venous Pressure

The venous pressure is maintained by:

(1) *Suction* as the *heart* expands.
(2) *Suction* due to *inspiration*.
(3) *Outside pressure* on the veins from the contraction of *muscles* and other organs, which massages the veins from without.

The force of suction is the most important factor here, and the main cause of the satisfactory return of venous blood to the heart. Where its force is interfered with, as in prolonged expiratory effort—e.g. in trumpet blowing—or from heart failure, venous return is impaired and congestion occurs. The massaging effect of muscle contraction and relaxation is very important, and it is for this reason that it is so much more exhausting to remain still in a fixed position than to move about. To maintain a position demanding violent muscular contraction, as does the model in some types of sculpture, or the porter or nurse who holds a part during an operation, will cause faintness in a comparatively short time.

15 The Lymphatic System

As BLOOD passes through the capillaries in the tissues fluid oozes out through their porous walls and circulates through the tissues themselves, bathing every single living cell. This fluid is called *tissue* or *interstitial fluid*; it fills the interstices or the spaces between the cells which form the different tissues. It is clear, watery, straw-coloured fluid similar to the plasma of the blood from which it is derived. While blood circulates only through the blood vessels, tissue fluid circulates through the actual tissue and carries food, oxygen, and water from the blood stream to each individual cell and carries away its waste products such as carbon dioxide, urea, and water, transmitting them to the blood. It is, in other words, the carrying medium between the tissue cells and the blood.

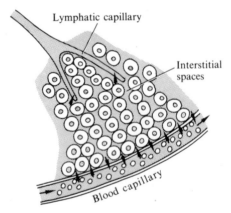

FIG. 119. Diagram showing the circulation of tissue fluid. Fluid passes out into the tissues and is collected partly by the blood capillary and partly by the lymph capillary.

Of the fluid which escapes from the capillaries into the tissues a certain amount passes back through the capillary wall, but its return is more difficult than its escape owing to the constant stream of oncoming blood which fills the capillaries. The excess fluid which cannot return directly into the blood stream is collected up and returned to the blood by a second set of vessels, which form the lymphatic system, and the fluid that these vessels contain is called lymph.

The *lymphatic system* consists of:

(1) Lymphatic capillaries.
(2) Lymphatic vessels.
(3) Lymphatic nodes.
(4) Lymphatic ducts.

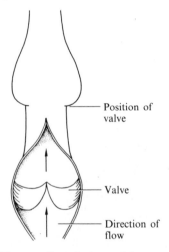

Position of valve

Valve

Direction of flow

FIG. 120. Lymphatic vessel (cut open).

The *lymphatic capillaries* arise in the spaces in the tissues as fine hair-like vessels with porous walls. These gather up excess fluid from the tissues and unite to form the lymphatic vessels.

The *lymphatic vessels* are thin-walled, collapsible tubes similar in structure to the veins, but carrying lymph instead of blood. They are finer and more numerous than the veins, and, like them, are provided with valves to prevent the lymph moving in the wrong direction. Lymphatic vessels are found in all tissues except the central nervous system, but run particularly in the subcutaneous tissues and pass through one or more lymphatic nodes.

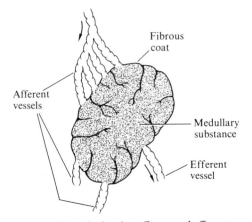

FIG. 121. A lymphatic node showing afferent and efferent vessels.

The Lymphatic Nodes

The lymphatic nodes are small bodies varying in size from a pinhead to an almond. Lymphatic vessels bring lymph to them and are called afferent vessels. These enter and divide up within the node and discharge the lymph into its substance. The lymph is then gathered up again into fresh lymphatic vessels called efferent vessels, which carry it on and ultimately empty it into the lymphatic ducts after possibly carrying the lymph through more nodes. The lymphatic nodes consist essentially of cells similar to the white blood corpuscles (lymphocytes), held together by a network of connective tissue, which also forms a capsule for the node.

The *functions of the lymphatic nodes* are:

(1) To *filter* the lymph of bacteria as it passes through. Thus when the tissues are infected the node may become swollen and tender. If the infection is a mild one, the organisms will be overcome by the cells of the node and the tenderness and swelling subside. If it is severe the organisms will cause acute inflammation and destruction of the white cells may result, causing an abscess to form in the node. If the bacteria

are not destroyed by the node they may pass on in the lymph stream and infect the general circulation, causing septicaemia.

(2) To provide fresh lymphocytes for the blood stream. The cells of the node constantly multiply and the newly-formed cells are carried away in the lymph.

(3) To produce some antibodies and antitoxins to prevent infection.

The lymphatic nodes are for the most part massed together in groups in various parts of the body. Groups in the neck and

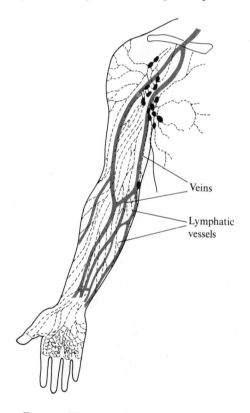

Veins

Lymphatic vessels

FIG. 122. The lymphatic vessels of the upper limb.

under the chin filter the lymph from the head, tongue, and the floor of the mouth. A group in the axilla filters the lymph from the upper limb and the chest wall. A group in the groin filters lymph from the lower limb and lower abdominal wall. Groups within the thorax and abdomen filter the lymph from the internal organs.

Special areas where much lymphatic tissue is found include: the palatine and pharyngeal tonsils; the thymus gland;

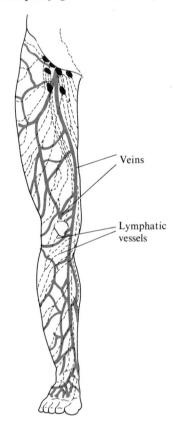

Fig. 123. The lymphatic vessels of the lower limb.

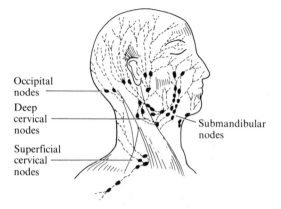

Occipital
nodes

Deep
cervical
nodes

Superficial
cervical
nodes

Submandibular
nodes

FIG. 124. The lymphatic vessels of the head and neck.

aggregated lymphatic follicles in the small intestine; the appendix; and the spleen.

The Spleen

The spleen is a large nodule of lymphoid tissue. By function it belongs to the circulatory system, as do the lymphatic nodes. It is a deep purplish-red colour, and lies high up at the back of the abdomen on the left side behind the stomach. It is enclosed in a capsule of fibrous tissue, and fibrous strands make a supporting meshwork throughout the gland. The spaces of this meshwork are filled in with a pulp-like material, called the splenic pulp, which is the essential substance of the organ and contains cells of different types. Many of these are similar to the lymphocytes of the blood and lymph nodes, and these help to produce fresh white cells for the blood stream. Others are phagocytes or devouring cells which engulf the red blood cells that are beginning to wear out and break them down.

The *functions of the spleen* are not fully known, but it is thought to be:

(1) A source of fresh lymphocytes for the blood stream.
(2) A seat for the destruction of *red blood cells*.
(3) The spleen is also thought to assist in fighting infection, as it becomes enlarged in certain diseases where the blood is infected, e.g. malaria and typhoid fever. It probably helps in

the manufacture of antibodies to fight infection. It is not essential to life, and can be removed by operation when it is responsible for ill-health, for example in haemolytic anaemia.

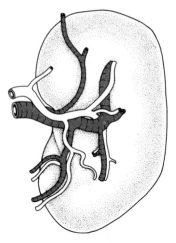

Fig. 125. The spleen and its blood vessels.

The Lymphatic Ducts

After filtration by the nodes the lymph is emptied by the lymphatic vessels into the two lymphatic ducts, the thoracic duct and the right lymphatic duct.

The *thoracic duct* is the larger. It begins in a small pouch at the back of the abdomen called the cisterna chyli. Into this empty all the lymphatic vessels from the lower limbs and the abdominal and pelvic organs. From the cisterna chyli the duct runs up through the mediastinum behind the heart to the root of the neck, where it turns to the left, is joined by the lymphatic vessels from the left side of the head and thorax and the left upper limb, and finally empties into the *left subclavian vein* at its junction with the left internal jugular vein. It is about 45 cm long and is provided with valves to prevent the lymph flowing in the wrong direction.

The *right lymphatic duct* is a comparatively small vessel formed by the joining of the lymphatic vessels from the right side of the head and thorax and the right upper limb at the root of the neck. It is only about 1 cm long and empties into the *right subclavian vein*, where it joins the right internal jugular vein.

The lymphatic ducts thus gather up all the lymph and return it to the blood stream, from which the fluid in the tissues is constantly renewed.

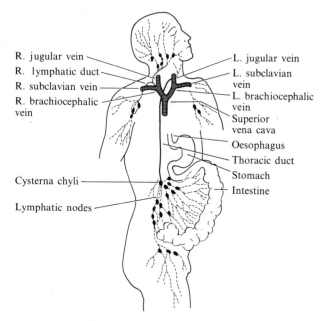

R. jugular vein
R. lymphatic duct
R. subclavian vein
R. brachiocephalic vein

L. jugular vein
L. subclavian vein
L. brachiocephalic vein
Superior vena cava
Oesophagus
Thoracic duct
Stomach
Intestine

Cysterna chyli
Lymphatic nodes

FIG. 126. The lymphatic ducts.

The Functions of the Lymphatic System

The functions of the lymphatic system are as follows:

(1). The lymphatic vessels gather up the excessive fluid or lymph from the tissues and thus permit a constant stream of fresh fluid to circulate through them.

(2) It is the channel by which excess proteins in the tissue fluid pass back into the blood stream.

(3) The nodes filter the lymph of bacterial infection and harmful substances.

(4) The nodes produce fresh lymphocytes for the circulation.

(5) The lymphatic vessels in the abdominal organs assist in the absorption of digested food, especially fat.

The Mechanism of the Lymphatic Circulation

The lymphatic circulation is maintained partly by suction, partly by pressure. Suction is the more important factor. The lymphatics empty into the large veins approaching the heart, and here there is negative pressure due to suction as the heart expands, and also suction towards the thorax during the act of inspiration.

Pressure is exerted on the lymphatics, as on the veins, by the contraction of muscles, and this outside pressure drives lymph onwards because the valves prevent a backward flow. There is also a slight pressure from the fluid in the tissues, on account of the constant pouring out of fresh fluid from the capillaries. If there is obstruction to the flow of the lymph through the lymphatic system, oedema results, i.e. a swelling of the tissues due to excess fluid collecting in them. The condition also results from any obstruction of the veins, since these also drain fluid from the tissues.

16 The Respiratory System

ALL LIVING cells require a constant supply of oxygen in order to carry on their metabolism. Oxygen is in the air, and the respiratory system is constructed in such a way that air can be taken into the lungs, where some of the oxygen is extracted for use by the body and at the same time carbon dioxide and water vapour are given up. The organs of the respiratory system are:

The nose
The pharynx
The larynx } leading to the lungs
The trachea
The bronchi
The bronchioles } within the lungs
The alveolar ducts and alveoli

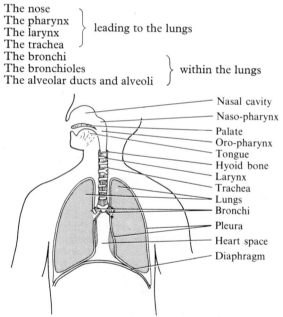

FIG. 127. Diagram of the respiratory tract.

The Nose

The external nose is the visible part of the nose, formed by the two nasal bones and by cartilage. It is both covered and lined by skin and inside there are hairs which help to prevent foreign material from entering. The nasal cavity is a large cavity divided by a septum. The anterior nares are the openings which lead in from without and the posterior nares are similar openings at the back, leading in to the pharynx. The roof is formed by the ethmoid bone at the base of the skull and the floor by the hard and soft palates at the roof of the mouth. The lateral walls of the cavity are formed by the maxilla, the superior and middle nasal conchae of the ethmoid bone and the inferior nasal concha. The posterior part of the dividing septum is formed by the perpendicular plate of the ethmoid bone and by the vomer, while the anterior part is made of cartilage.

The three nasal conchae project into the nasal cavity on each side and greatly increase the surface area of the inside of the nose. The cavity of the nose is lined throughout with ciliated mucous membrane, which is extremely vascular, and atmospheric air is warmed as it passes over the epithelium, which contains many capillaries. The mucus moistens the air and entraps some of the dust and the cilia move the mucus back into the pharynx for swallowing or expectoration. The nerve endings of the sense of smell are situated in the highest part of the nasal cavity round the cribriform plate of the ethmoid bone.

Some of the bones surrounding the nasal cavity are hollow. The hollows in the bones are called the *paranasal sinuses*, which both lighten the bones and act as sounding chambers for the voice, making it resonant. The maxillary sinus lies below the orbit and opens through the lateral wall of the nose. The frontal sinus lies above the orbit towards the midline of the frontal bone. The ethmoidal sinuses are numerous and are contained within the part of the ethmoid bone separating the orbit from the nose, and the sphenoidal sinus is in the body of the sphenoid bone. All the paranasal sinuses are lined with mucous membrane and all open into the nasal cavity, from which they may become infected (see Fig. 44, page 69).

The Pharynx

The roof of the pharynx is formed by the body of the sphenoid bone and inferiorly it is continuous with the oesophagus. At the back it is separated from the cervical vertebrae by loose connective tissue, while the front wall is incomplete and communicates with the nose, mouth and larynx. The pharynx is divided into three sections, the naso-pharynx which lies behind the nose, the oro-pharynx which lies behind the mouth, and the laryngeal pharynx which lies behind the larynx.

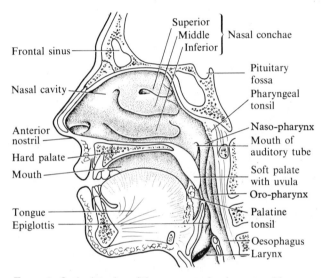

FIG. 128. Sagittal section of the nose, mouth, pharynx and larynx.

The naso-pharynx is that part of the pharynx which lies behind the nose above the level of the soft palate. On the posterior wall there is a patch of lymphoid tissue called the pharyngeal tonsil, commonly referred to as adenoids. This tissue sometimes enlarges and blocks the pharynx causing mouth breathing in children. The auditory tubes open from the lateral walls of the naso-pharynx and through them air is carried to the middle ear. The naso-pharynx is lined with ciliated mucous membrane which is continuous with the lining of the nose.

The oro-pharynx lies behind the mouth below the level of the soft palate, with which its lateral walls are continuous.

Between the folds of these walls, which are called the palato-glossal arches, are collections of lymphoid tissue called the palatine tonsils. The oro-pharynx is part of both the respiratory tract and the alimentary tract, but it cannot be used for swallowing and breathing simultaneously. During the act of swallowing breathing stops momentarily and the oro-pharynx is completely blocked off from the naso-pharynx by the raising of the soft palate. The oro-pharynx is lined with stratified epithelium.

The Larynx

The larynx is continuous with the oro-pharynx above and with the trachea below. Above it lie the hyoid bone and the root of the tongue. The muscles of the neck lie in front of the larynx, and behind the larynx lies the laryngo-pharynx and the cervical vertebrae. On either side are the lobes of the thyroid gland. The larynx is composed of several irregular cartilages joined together by ligaments and membranes.

The *thyroid cartilage* is formed of two flat pieces of cartilage fused together in the front to form the laryngeal prominence

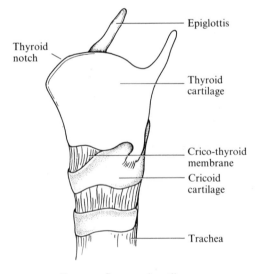

FIG. 129. Laryngeal cartilages.

or Adam's apple. Above the prominence is a notch called the thyroid notch. The thyroid cartilage is larger in the male than in the female. The upper part is lined with stratified epithelium and the lower part with ciliated epithelium.

The *cricoid cartilage* lies below the thyroid cartilage and is shaped like a signet ring with the broad portion at the back. It forms the lateral and posterior walls of the larynx and is lined with ciliated epithelium.

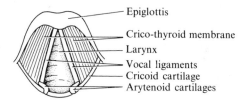

Epiglottis
Crico-thyroid membrane
Larynx
Vocal ligaments
Cricoid cartilage
Arytenoid cartilages

Vocal ligaments resting

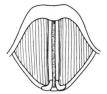

Vocal ligaments during speech

FIG. 130. Diagram of vocal ligaments.

The *epiglottis* is a leaf-shaped cartilage attached to the inside of the front wall of the thyroid cartilage immediately below the thyroid notch. During swallowing the larynx moves upwards and forwards so that its opening is occluded by the epiglottis.

The *arytenoid cartilages* are a pair of small pyramids made of hyaline cartilage. They are situated on top of of the broad part of the cricoid cartilage and the vocal ligaments are attached to them. They form the posterior wall of the larynx.

The hyoid bone and the laryngeal cartilages are joined together by ligaments and membranes. One of these, the *cricothyroid membrane*, is attached all round to the upper edge of the cricoid cartilage and has a free upper border, which is not circular like the lower border, but makes two parallel lines running from front to back. The two parallel edges are the *vocal ligaments*. They are fixed to the middle of the thyroid cartilage in front and to the arytenoid cartilages behind, and they contain much elastic tissue. When the intrinsic muscles of the larynx alter the position of the arytenoid cartilages the vocal ligaments are pulled together, narrowing the gap between them. If air is forced through the narrow gap, called the chink, during expiration, the vocal ligaments vibrate and sound is produced. The pitch of the sound produced depends on the length and tightness of the ligaments; an increased tension gives a higher note, a slacker tension a lower note. Loudness depends on the force with which the air is expired. The alteration of the sound into different words depends on the movements of the mouth, tongue, lips and facial muscles

The Trachea

The trachea begins below the larynx and runs down the front of the neck into the chest. It divides into the right and left main bronchi at the level of the fifth thoracic vertebra. It is about 12 cm long. The isthmus of the thyroid gland crosses in front of the upper part of the trachea, and the arch of the aorta lies in front of the lower part, with the manubrium of the sternum in front of it. The oesophagus lies behind the trachea, separating it from the bodies of the thoracic vertebrae. On either side of the trachea lie the lungs, with the lobes of the thyroid gland above them. The wall of the trachea is made of involuntary muscle and fibrous tissue strengthened by incomplete rings of hyaline cartilage. The deficiency in the cartilage lies at the back where the trachea is in contact with the oesophagus. When a bolus of food is swallowed the oesophagus is able to expand without hindrance, but the cartilage maintains the patency of the airway. The trachea is lined with ciliated epithelium containing goblet cells which secrete mucus. The cilia sweep the mucus and foreign particles upwards towards the larynx.

The Lungs

The lungs are two large, spongy organs lying in the thorax on either side of the heart and great vessels. They extend from the root of the neck to the diaphragm and are roughly cone-shaped with the apex above and the base below. The ribs, costal

cartilages and intercostal muscles lie in front of the lungs
and behind them are the ribs, the intercostal muscles and the
transverse processes of the thoracic vertebrae. Between the
lungs is the *mediastinum*, a block of tissue which completely
separates one side of the thoracic cavity from the other,
stretching from the vertebrae behind to the sternum in front.
Within the mediastinum lie the heart and great vessels, the
trachea and the oesophagus, the thoracic duct and the thymus
gland. The lungs are divided into lobes. The left lung has two
lobes, separated by the oblique fissure. The superior lobe is
above and in front of the inferior lobe which is conical in shape.
The right lung has three lobes. The inferior lobe is separated
by an oblique fissure in a similar manner to the left inferior
lobe. The remainder of the lung is separated by a horizontal
fissure into the superior lobe and the middle lobe. Each lobe is
further divided into named broncho-pulmonary segments,
separated from each other by a wall of connective tissue and
each having an artery and a vein. Each segment is also divided
into smaller units called lobules.

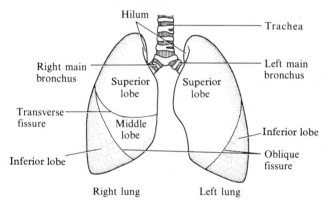

FIG. 131. Diagram showing lobes of lungs.

The Bronchi

The two main bronchi commence at the bifurcation of the
trachea and one leads into each lung. The left main bronchus is
narrower, longer and more horizontal than the right main
bronchus because the heart lies a little to the left of the mid-

line. Each main bronchus divides into branches, one for each lobe. Each of these then divides into named branches, one for each broncho-pulmonary segment, and divides again into progressively smaller bronchi within the lung substance. The bronchi are similar in structure to the trachea but the cartilage is less regular.

The Bronchioles

The finest bronchi are called bronchioles. They have no cartilage but are composed of muscular, fibrous and elastic tissue lined with cuboid epithelium. As the bronchioles become smaller, the muscular and fibrous tissue disappears and the smallest tubes, called terminal bronchioles, are a single layer of flattened epithelial cells.

The Alveolar Ducts and Alveoli

The terminal bronchioles branch repeatedly to form minute passages called alveolar ducts, from which alveolar sacs and alveoli open. The alveoli are surrounded by a network of capillaries. Deoxygenated blood enters the capillary network from the pulmonary artery and oxygenated blood leaves it to enter the pulmonary veins. It is in the capillary network that the change of gases takes place between the air in the alveoli and the blood in the vessels.

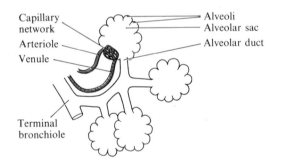

FIG. 132. Diagram of alveoli.

The Hilum of the Lung

The hilum is a triangular shaped depression on the concave medial surface of the lung. The structures forming the root of the lung enter and leave at the hilum, which is at the level of the fifth to seventh thoracic vertebrae. These structures include the main bronchus, the pulmonary artery, the bronchial artery and branches of the vagus nerve, which enter at this point, and two pulmonary veins, the bronchial veins and lymphatic vessels, which leave the lung at the root. There are also many lymph nodes round the root of the lung.

The Pleura

The pleura is a serous membrane which surrounds each lung. It is composed of flattened epithelial cells on a basement membrane and it has two layers. The visceral pleura is firmly attached to the lungs, covering their surfaces and dipping into the inter-lobar fissures. At the root of the lungs the visceral layer is reflected back to become the parietal layer which lines the chest wall and covers the superior surface of the diaphragm. The two layers of the pleura are normally in close contact with each other, separated only by a film of serous fluid which enables them to glide over one another without friction. This potential space between the two layers is called the pleural cavity.

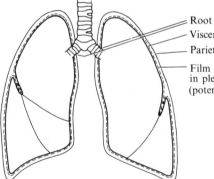

Root of lungs
Visceral pleura
Parietal pleura
Film of serous fluid in pleural cavity (potential space)

FIG. 133. Diagram of pleura.

Gaseous Exchange

Exchange of gases within the body takes place both in the lungs, where it is called external respiration, and in the tissues, where it is called internal respiration. An elementary law of physics relating to gases states that gases tend to diffuse from a higher pressure to a lower pressure. The air which is breathed in contains several gases. The composition of inspired air is:

> Nitrogen 79 per cent
> Oxygen 21 per cent
> Carbon dioxide 0·04 per cent
> Water vapour
> Traces of other gases

External respiration. When inspired air reaches the alveoli it is in close contact with the blood in the surrounding capillary network. Oxygen in the alveoli, at a pressure of 100 mmHg, comes in contact with oxygen in the venous blood at a pressure of 40 mmHg and therefore the gas diffuses into

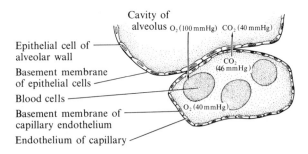

Cavity of
alveolus O_2 (100 mmHg) CO_2 (40 mmHg)

Epithelial cell of alveolar wall

Basement membrane of epithelial cells

CO_2 (46 mmHg)

Blood cells

Basement membrane of capillary endothelium

O_2 (40 mmHg)

Endothelium of capillary

FIG. 134. Diagram of gaseous exchange in the lungs.

the blood until the pressures are equal. At the same time carbon dioxide in the blood, at a pressure of 46 mmHg, comes in contact with alveolar carbon dioxide at a pressure of 40 mmHg and therefore the gas diffuses out of the blood into the alveoli. The gaseous content of expired air is therefore altered to contain less oxygen and more carbon dioxide. The nitrogen content remains the same. The composition of expired air is:

> Nitrogen 79 per cent
> Oxygen 16 per cent
> Carbon dioxide 4·5 per cent
> Water vapour
> Traces of other gases

Internal respiration. The oxygen which has diffused into the blood is carried in the haemoglobin—now called oxy-haemoglobin— to the tissues. Here the pressure of the oxygen is low, so the gas diffuses out of the blood and into the tissues, the amount depending on the activity in the tissue. At the same time carbon dioxide, produced in the tissues, is carried away by the blood.

Mechanism of Respiration

Respiration consists of two parts, inspiration and expiration. The chest expands during inspiration owing to movement of the diaphragm and the intercostal muscles. When the diaphragm contracts during inspiration it is flattened and lowered and the thoracic cavity is increased in length (see Chapter 11). The external intercostal muscles, on contraction, lift the ribs and draw them out, increasing the depth of the thoracic cavity. As the chest wall moves up and out the parietal pleura, which is closely attached to it, moves with it. The visceral pleura follows the parietal pleura and the volume of the interior of the thorax is increased. The lung expands to fill the space and air is sucked into the bronchial tree.

Expiration during quiet breathing is passive. The diaphragm relaxes and assumes its original domed shape. The intercostal muscles relax and the ribs revert to their previous position. The lungs recoil and air is driven out through the bronchial tree. In forced expiration the internal intercostal muscles contract actively to lower the ribs. The accessory muscles of respiration may be brought into use during deep breathing or when the airway is obstructed. During inspiration the sternocleidomastoid muscles raise the sternum and increase the diameter of the thorax from front to back. Serratus anterior and pectoralis major pull the ribs outwards when the arm is fixed. Latissimus dorsi and the muscles of the anterior abdominal wall help to compress the thorax during forced expiration.

Control of Respiration

Respiration is controlled by the respiratory centre in the medulla oblongata. The presence of an accumulation of carbon dioxide in the blood stimulates specialized cells in the great arteries. Impulses are carried by the vagus and glosso-pharyngeal nerves to the respiratory centre, from the respiratory centre by the phrenic nerves to the diaphragm and by the intercostal nerves to the intercostal muscles. These impulses cause the muscles to contract and inspiration occurs.

The Capacity of the Lungs

Tidal volume is the amount of air breathed in and out during normal quiet breathing (about 500 ml). After a normal expiration the amount of air taken in during forced inspiration is called the *inspiratory capacity* (about 3000 ml). It includes the tidal volume. After a quiet expiration it is possible to force another 1000 ml of air from the lungs. This is the *expiratory reserve volume*.

Table 10
Lung Capacity

Inspiratory reserve volume (2500 ml) ⎫	Inspiratory capacity (3000 ml)	⎫
Tidal volume (500 ml) ⎬		
Expiratory reserve volume (1000 ml)		Vital capacity (4000 ml)
Residual volume (1100 ml)		

After the deepest possible expiration there is still the residual volume left in the respiratory passages. This is called the *residual volume* (about 1100 ml).

Vital capacity is the largest volume of air expired after the deepest possible inspiration (about 4000 ml). An important measure of respiratory function is the amount of forced expiration in one second. It is called the *forced expiratory volume* and should measure 75 to 80 per cent of the vital capacity.

17 Nutrition

GOOD health is dependent on satisfactory nutrition, and this is in turn dependent on plentiful supplies of the foods which are necessary to healthy life. Food is one of the essential needs of the body. Substances which can serve as food for the body are those which it can use as *fuel* for combustion, or *building material* for the repair and growth of the tissues. Fuel is required to produce the energy for the activities of every living thing and to maintain the heat at which the individual exists. Building materials are necessary to repair the body tissues, since they are constantly active and being worn out by their activities; in addition, in infants and children, extra building material is required to build up the new tissues needed for the processes of growth. Fuel supplies and building materials alone are, however, not enough. Certain other substances are necessary to enable the tissues to use the building materials and the fuel supplies, and these are known as vitamins.

There are six essential foodstuffs with which the body must be constantly supplied through the foods that we eat. These are:

- (1) Proteins.
- (2) Carbohydrates.
- (3) Fats.
- (4) Water.
- (5) Mineral salts.
- (6) Vitamins.

Every article of food contains one or more of these foodstuffs and is only of value as food because it does contain them. The proteins, water, and salts are the body-building foods; the carbohydrates and fats are essentially the fuel foods, though the body can and does use protein also as fuel, if more is taken in than is required for body building, or if there is a lack of other fuel, as in starvation. The vitamins and certain salts act as regulators of tissue activity, so that, though vitamins are of no use either as fuel or as body builders, the nutrition of the tissues suffers if they are not present in sufficient quantities in the food supply, and diseases appear which can be both prevented and cured by ensuring that the vitamins are present in satisfactory quantities in the diet.

To be of use to the body these foodstuffs must be digested and absorbed. Food must therefore be of such a nature that it can be digested, i.e. broken down by digestive juices into substances that can pass into the blood stream, and be carried to the various tissues for their use. Proteins, carbohydrates and fats are complicated compounds found in plant and animal matter, and require digestion. Water and mineral salts are simple inorganic substances; they can be absorbed, therefore, without digestion, and they enter into the composition of all animal and plant matter. In fact, all living matter consists largely of water. The inorganic salts, absorbed from the soil or water, are built up by living things into organic salts which are an essential part of all plants and animals. Before studying the digestive system in detail, it is necessary to understand the nature of the different foods and the meaning of the term digestion.

Proteins are the most complicated of the foodstuffs. They consist of carbon, hydrogen, oxygen, nitrogen and sulphur and usually phosphorus. They are often spoken of as the nitrogenous foodstuffs, as they are the only ones which contain the element nitrogen. They are essential for the building up of living protoplasm, as this also consists of the elements carbon, hydrogen, oxygen, nitrogen and sulphur. They are found in both animal and plant matter, but the animal proteins are the most valuable to the human body as building material, since they are similar to human protein in composition. On the other hand, plant proteins are cheaper; they are more useful as body fuel than as body builders, but do provide at a lower cost some of the amino acids that the body needs for tissue building.

The **sources of protein** are:

(1) *Animal:*
    (a) Eggs, containing albumin.
    (b) Lean meat, containing myosin.
    (c) Milk, containing caseinogen and lactalbumin.
    (d) Cheese, containing casein.

(2) *Plant:*
    (a) Wheat and rye, containing gluten.
    (b) Pulse foods (e.g. peas and beans) containing legumin.

All proteins are made up of simpler substances known as *amino acids*. There are about twenty of these amino acids, but each protein contains some of these only. The amino acids are like letters from which many words can be made, each word being a different combination of letters. The protein of each different type of animal or plant is a different combination of amino acids. Ten essential amino acids are found in human protein. These are amino acids which the body cannot build up for itself. Proteins which contain all ten are called *complete proteins*, e.g. albumin, myosin, casein. Proteins which do not contain all the ten are called *incomplete proteins*, e.g. gelatin, which is contained in all fibrous tissue, and is extracted from bone and calves' feet in the making of soups and jelly. Animal proteins, such as those of eggs, milk and meat, not only contain all the ten amino acids the body needs, but contain all of them in good proportion; these are called *first-class proteins* and are the best building material for the body tissues. Plant proteins, such as gluten and legumin, contain only slight quantities of one or more of the ten amino acids essential to the body, and are therefore called *second-class proteins*, as they äre not such good building material. Some first-class animal protein should always be included in the diet.

Carbohydrates include *sugar* and *starch*. They consist of carbon, hydrogen, and oxygen, and always contain hydrogen and oxygen in the same proportion as in water, i.e. twice as much hydrogen as oxygen. They are the chief sources of body fuel, being most easy to digest and absorb and most readily burnt in the tissues, being broken down into carbon dioxide and water. They are obtained chiefly from plant foods.

The *sources of starch* are:

(1) Cereals, e.g. wheat, rice, barley.
(2) Tubers and roots, e.g. potatoes, parsnips.
(3) Pulse foods, e.g. peas, beans, lentils.

The *sources of sugar* are:

(1) The sugar cane (sucrose).
(2) Beetroot and all sweet vegetables and fruits, e.g. grape sugar or glucose.
(3) Honey.
(4) Milk, containing lactose or milk sugar.

Sugars are of three types:

(1) Simple sugars (monosaccharides), such as glucose or grape sugar (formula, $C_6H_{12}O_6$).

(2) Complex sugars (disaccharides), such as cane sugar or sucrose, and milk sugar or lactose (formula, $C_{12}H_{22}O_{11}$).

(3) Polysaccharides—the most complex carbohydrates; starches such as potatoes, cereals and root vegetables.

Starch differs from sugar in that it is insoluble in water. Plants store sugar in the form of starch to prevent it escaping in solution into the water in the soil in which they live. (Formula for starch, $n(C_6H_{10}O_5)$, a polysaccharide; n stands for different numbers in the different starches of various plants.) All carbohydrates are reduced to monosaccharides before they can be absorbed from the digestive tract.

Fats, like carbohydrates, consist of carbon, hydrogen, and oxygen, but do not contain so much oxygen in proportion to the hydrogen present. They also serve as body fuel. They are the best source of fuel from the point of view that 1 g of fat produces more than twice as much energy as 1 g of sugar. On the other hand, they are not so easy to digest and absorb, and not so readily burnt in the tissues. Fats can only be completely burnt to form the final products of combustion, carbon dioxide and water, if they are burnt with sugar. If sufficient sugar is not burnt with them, the combustion of fat is incomplete and acid or acetone bodies are formed in the tissues. These acetone bodies cause fatigue in the muscles, and if present in large quantities alter the reaction of the blood, causing a condition known as acidosis, which may lead to coma and death. Severe acidosis is only likely to occur in diabetic patients, who are unable to burn sugar, and in starvation, when the small quantity of sugar which the body can store has been used up and the large quantities of fat which the body can store are forming the main source of body fuel.

Fats are obtained from both animal and plant matter. The chief *sources of fat* are:

(1) *Animal:*
    (a) Fat meat and fish oils.
    (b) Butter.
    (c) Milk and cream.

(2) *Plant:*
    (a) Nut oil, contained in margarine.
    (b) Olive oil.

Fats are compounds of glycerin and fatty acids. Different fats contain different acids, e.g. fat meat contains stearic acid, butter contains butyric acid, olive oil contains oleic acid, and so on.

Animal fats are more expensive than plant fats, but are more valuable as food because they contain the vitamins A and D, which are both fat-soluble, provided the animals have been exposed to sunlight and not kept in stalls. However, plant fat can be made equally valuable by exposing it to the action of ultraviolet rays, and all margarine, whether manufactured from fish oil or plant oil, is so treated today, to ensure the supply of these vitamins. In this way the population has been protected from deficiency diseases due to lack of these fat-soluble vitamins.

Water is a simple compound of hydrogen, 2 parts and oxygen, 1 part. It forms two-thirds of the body and is present in most of the foods we eat. Lean meat is three-quarters water, milk contains 87 per cent water, while cabbage contains as much as 92 per cent water. In addition to the water contained in food the body needs 2 to 3 litres of water every day. Water is required for many purposes, the chief being:

(1) The building of body tissues and body fluids.
(2) The excretion of waste products.
(3) The making of digestive and lubricating fluids.
(4) The cooling of the body by the evaporation of sweat.

Water forms 60 per cent of the body weight; 70 per cent of the water is contained in the body tissues and is called the *intracellular* fluid; 30 per cent forms the *extracellular* fluid, the tissue or interstitial fluid forming 15 to 20 per cent, and the blood plasma forming 10 to 15 per cent. Water forms 96 per cent of the urine (1·5 litres is the normal daily volume) and a large proportion of the sweat (500 to 600 ml is the average volume); the lungs lose 250 to 350 ml of water and the faeces 100 to 150 ml. Some 8 to 10 litres of water are secreted into the mouth and alimentary canal in the various digestive juices. This is largely reabsorbed, leaving only a small quantity in the faeces. Consideration of these figures shows how important water is among the foodstuffs.

Salts are produced by the action of an acid on a mineral, which is termed the base of the salt, e.g. sodium chloride or common salt, produced by the action of hydrochloric acid on sodium; calcium lactate, produced by the action of lactic acid on calcium. Various different salts and the ions split off from them are required by the body, every tissue and fluid in the body containing them. Chlorides, carbonates, and phosphates of sodium, potassium, and calcium are particularly important.

Salts are required for body building and are also regulators of tissue activity. They provide the electrolytes, capable of carrying positive and negative electric charges in the body fluids (see Chapter 1). A correct balance of electrolytes is essential for normal activity in the body tissues and fluids.

The sodium, potassium and calcium ions provide the positively-charged ions, and the chlorine, carbonate and phosphate ions produce the negatively-charged ions, the two balancing one another and providing the almost neutral reaction necessary for the life of the tissues.

Sodium is present in all tissues; the body contains it in the form of sodium chloride to the extent of 9 g to 1 litre (0·9 per cent). It is present in the same concentration in all tissue fluid. Sodium carbonate and sodium phosphate are also always present in the blood and tissues. The carbonates give the alkaline reaction to the blood and form the alkali reserve, neutralizing the carbonic acid produced by fuel combustion. The phosphate is the carrier of the acids produced by the breaking down of the body building foods and carries them to the kidneys, by which they are excreted. It is obtained from our food, particularly animal foodstuffs, and from the rock salt used in cooking.

Potassium is present in all tissue cells, where it replaces the sodium of blood and tissue fluid as the base and the source of the negatively charged ions. It is obtained from our food, particularly from plant foodstuffs.

Calcium is present in all tissues, particularly in bone, in teeth and in blood, and is necessary for normal functioning of nerves. It is obtained chiefly from milk, cheese, eggs and green vegetables, and to some extent from hard water, though this inorganic form is less valuable than the organic salts in plant and animal foods. Adults require 400 to 500 mg daily.

Iron is essential for the formation of the haemoglobin of red blood cells. It is obtained from green vegetables, particularly spinach and cabbage, egg yolk and red meats. Men require 10 mg daily; women 10 to 15 mg.

Phosphorus is also needed for the building of the body tissues. It is obtained from yolk of egg, milk and green vegetables.

Iodine is required for the formation of the secretion of the thyroid gland. It is obtained from seafood and is present in green vegetables, which have derived it from the fine spray of sea water in the air which is carried inland by winds.

Calcium, iron and iodine are the only minerals likely to be insufficient. The others are present in adequate amounts in the diet.

Vitamins

Vitamins are substances which are also essential to normal health, although they are of no value to the body as fuel nor as building material. Without them diseases occur which are known as deficiency diseases. They are present in small quantities in living foodstuffs and are required in minute traces only each day. Their discovery dates back to the period following the 1914–18 war, and as their composition was at first unknown they were named after the letters of the alphabet. They can today be made synthetically to a large extent.

There are numbers of vitamins now identified, but experimental work still continues. The chief are: vitamin A, the vitamin B complex, vitamin C, vitamin D, vitamin E and vitamin K.

Vitamin A is present in animal fats, being fat-soluble. In carrots and green vegetables, and in all yellow fruits a substance *carotene* is found, which, like the green colouring matter chlorophyll, is formed under the influence of sunlight. This is a precursor of vitamin A, and animals, including man, can turn the carotene present in their food into vitamin A within their bodies. Lack of vitamin A causes stunted growth and lowered resistance to infection. The mucous membranes particularly are unhealthy and an easy prey to bacteria when the diet is deficient in it. The conjunctiva of the eye is affected and a form of conjunctivitis occurs, known as *xerophthalmia*, in which the conjunctiva loses its transparency, and a horniness or cornification, develops in it. The retina is also affected and night blindness (i.e. inability to see at night) develops—a symptom which became apparent in the 'black-out' during the war, when some people found they could not see as well as others in the dark, because their diet had not been satisfactory.

The **vitamin B** complex consists of a number of factors, although originally thought to be a single substance. These factors were confused because they had a similar distribution, being found particularly in the husks and germ of cereals and pulses, and in yeast and yeast extracts. They are also present to a lesser extent in vegetables, fruit, milk, eggs and meat. White flour, and the bread, cakes and pastries made from it, do not contain it, nor does polished rice and barley. Hence brown bread and wholemeal flour have a food value that white bread and flour lack, and where these latter form a large part of the diet a subnormal state of health may be present, or even deficiency disease, as occurred in many of the prison camps during the last war, especially in the Far East.

The chief factors in the vitamin B complex are:

Vitamin B₁ (*aneurine* or *thiamine*). This is essential for carbohydrate metabolism and controls the nutrition of nerve cells and a marked deficiency of it leads to beri-beri, in which there is inflammation of the nerves, causing paralysis and a loss of tone and activity in the intestinal muscles, with constipation, while the patient complains of loss of appetite and a burning sensation in the feet.

Vitamin B₂ (*riboflavine*). This is essential for proper functioning of cell enzymes.

Vitamin B₃ (*nicotinic acid*). This is essential for carbohydrate metabolism. A lack of this causes pellagra, which leads to skin eruptions, gastro-intestinal changes and mental changes.

Vitamin B₆ (*pyridoxine*). This is believed to be necessary for protein metabolism.

Vitamin B₁₂ (*cyanocobalamin*). This is the anti-anaemic substance or factor absorbed by the villi of the small intestine and stored in the liver. It is satisfactorily absorbed only in the presence of an intrinsic factor produced by the lining of the stomach and hydrochloric acid. Vitamin B₁₂ is essential for the proper development of red cells in the red bone marrow. Lack of it, or of its absorption, causes pernicious anaemia.

Folic acid. This is part of the vitamin B complex. It is also necessary for the maturation of red blood cells.

Vitamin C (ascorbic acid) is water-soluble and is found in fresh fruit, particularly citrus fruits (oranges, grapefruit and lemons) and in green vegetables and potatoes. It is important in tissue respiratory activity, wound repair and resistance to infection, and affects the condition of capillary walls, which become abnormally fragile if it is not present in plentiful supply in the diet. Lack of it causes scurvy, so that it is called the antiscorbutic vitamin. It is particularly readily destroyed by heat, hence some fresh fruit or salad should be included in the daily diet. Cabbage and other greens provide a very rich source of vitamin C and are comparatively cheap. Eaten raw, as salads, in a finely shredded form, they are very palatable and most valuable, provided the cabbage is fresh and crisp. Cooking for a long period of time lessens the content of vitamin C, and for this reason cabbage should be finely shredded and cooked in boiling water only till tender, i.e. for 10 to 15 minutes.

Under normal circumstances the eating of a good mixed

diet, containing plenty of fresh fruits and vegetables, makes the taking of vitamin C in tablet form unnecessary and inadvisable.

Vitamin D is fat-soluble and is found with vitamin A in animal fats, provided the animal has been in the sun. Cod-liver and halibut-liver oil are very rich in it. Halibut oil is more expensive as the supplies are less plentiful, but it is richer in content, so that it is useful for those who cannot digest cod-liver oil. Vitamin D can also be made in the skin by the action of ultraviolet rays on the ergosterols present. Further, it can be manufactured by subjecting the sterols in fats to the action of ultraviolet rays. The product of this process is named **calciferol**, and this can be obtained in tablet form. It is identical with the vitamin D present in animal fat and can be used in illness and in emergencies as a substitute for it. It is essential for the development of bone and teeth, affecting the use of calcium and phosphorus in the body. Lack of vitamin D causes rickets, so that it is known as the anti-rachitic vitamin. This disease was common among the poor of industrial areas who ate margarine in place of butter, especially in countries like the British Isles, where there was not a great deal of sunshine. However, modern methods of manufacturing the vitamin have made it possible to add it to margarine, with the result that rickets as a disease has practically disappeared in this country, since it has become compulsory for all makers of margarine to ensure that it contains the vitamin.

Vitamin E is present in vegetable oils. It is present in cereals and has been shown to be essential for reproduction in rats. Little is known of its importance in human beings.

Vitamin K is fat-soluble and can be obtained from green vegetables and liver. Persons who have had deficiency in their diet show a tendency to haemorrhage, since vitamin K helps to form the prothrombin in the blood. It is synthesized in the intestines by bacterial action.

The quantities of the various vitamins required in the normal diet are estimated in milligrams, pure vitamins can be given in measured quantities on the doctor's prescription.

Roughage

The food should also contain cellulose. This is indigestible and therefore remains in the bowel and stimulates it to empty itself. It is called roughage. Cellulose forms the fibrous part of plant foods and is present in all green vegetables, in fruits, the skins of peas and beans, and in wholemeal flour and bread. These foods are therefore essential to prevent constipation.

Diet

The diet is the daily ration of foods required by the individual. Man requires a mixed diet, i.e. a diet consisting of different animal and plant foods, as no single article of food contains all the essential foodstuffs in the proportions required for health. Milk and oysters contain all the nutrients or foodstuffs, but not in the proportions the body requires. Cow's milk contains on an average:

Protein (caseinogen, lactalbumin)	4 per cent
Lactose	4 to 5 per cent
Fats	3·5 per cent
Mineral salts	0·7 per cent
Water	87 to 88 per cent

A well-balanced diet should contain the foodstuffs roughly in the following proportions: 1 part protein, 1 part fat to 4 parts of carbohydrate. In addition it must contain traces of the various vitamins. The standard requirements per day are estimated by many authorities to be:

Protein	55 to 70 g
Fats	70 to 100 g or $\frac{1}{3}$ of total calorie intake
Carbohydrate	300 to 400 g or 50 to 60 per cent of total calorie intake

It is now accepted that health can be maintained on a diet containing 70 g of protein and 70 g of fat. These are the scarcer and more expensive foodstuffs and are derived largely from animals. Larger amounts of protein are required by persons carrying out heavy work, either manual labour or a hard athletic programme; also by pregnant mothers and growing children, especially in late adolescence. In the first group there is a quicker breaking down of tissue, and in the second there is the demand for building material to make new tissue for growth, in addition to the constant repair of cells that are worn out by the normal activities of life. More fat is especially desirable for persons living in a cold environment and when prolonged staying power is required, since it is more slowly digested and absorbed. Where a high calorie intake is necessary, about half the calories should be derived from fat, otherwise the meals become too bulky. In many of the under-developed and thickly populated countries, lack of animal protein and fat is responsible for a low standard of health and a high death rate, especially in the lower age groups.

The amount of carbohydrate required depends on the energy

output. A man doing heavy work requires more than a sedentary worker.

Caloric Value of Diet

The value of foods is reckoned by the amount of heat which
they yield on combustion. The heat is measured in Calories,*
now known as kilocalories. A Calorie is the amount of heat
required to raise the temperature of 1 litre of water 1°C. The
Caloric value of each of the foodstuffs is known and is as
follows:

> 1 g of protein has a heat value of 4 Calories.
> 1 g of carbohydrate has a heat value of 4 Calories.
> 1 g of fat has a heat value of 9 Calories.

The Caloric value of an average diet should be 2500 to 3000
Calories per day. A man doing active work will require 3300
Calories. A sedentary worker will require about 2850 Calories.
A woman will require about 2200 Calories.

The Caloric value required by the individual is affected by:

(1) *Age.* Children require more in proportion to their
weight, because they need food for growth as well as other
activities; a baby requires 50 Calories per pound. They need
more protein in proportion to fuel foods, because of their
rapid growth.

(2) *Exercise.*

(3) *Sex.* A male requires more than a female, the female
requiring four-fifths of the quantity the male requires. In the
16- to 18-year age group the male adolescent requires on an
average one-fifth more than the male adult and the female
adolescent the same quantity as the male adult; this is because
of the rapid growth and development that takes place at this
period, in addition to the energy expended in physical activity.

(4) *Weight and build.* The heavier the individual, the more
food is required, except where weight is due to fat.

(5) *Climate and weather.* In hot climates and hot weather a
diet of lower Caloric value is sufficient.

(6) *Temperament.* The placid person requires less food than
the excitable one.

Digestion

Digestion is the breaking down of the foodstuffs into simple
substances which can be absorbed. It is a chemical process. Of

---

* The word calorie, with a small 'c,' is the amount of heat required to raise the
temperature of 1 millilitre of water 1°C.

the essential foodstuffs, proteins, carbohydrates and fats are complicated compounds built up by plants and animals, and have large molecules which cannot be absorbed into the living cell. To make them absorbable they must be broken down into simpler compounds with a smaller molecule.

Digestion is brought about by *enzymes* contained in the various digestive juices secreted by various glands such as the salivary glands and the gastric glands. An enzyme is a substance which causes a chemical change in another substance without being changed itself. The digestive enzymes cause the various foodstuffs to combine with water, as a result of which the foodstuff is split up into simpler substances (this process is known as hydrolysis). For example, complex sugar such as sucrose or cane sugar combines with water and splits up into two molecules of simple sugar, which can be absorbed. This can be expressed by formulae as follows:

$$C_{12}H_{22}O_{11} + H_2O = C_{12}H_{24}O_{12} = 2(C_6H_{12}O_6)$$
(Cane sugar and water   (Glucose)

This is a very simple example; in other cases the process of combining with water and splitting must take place over and over again before a substance which the body can absorb is produced.

The enzymes which cause digestion act best at body temperature. Heat destroys them and cold checks their action. Each enzyme also acts in a specific reaction, either acid or alkaline. Each digestive juice therefore contains not only enzymes, but also acids or alkalis to give the required reaction, e.g. saliva is alkaline, but gastric juice is acid.

Each enzyme acts on one foodstuff only, either on proteins, carbohydrates, or fats. *Protein-splitting* enzymes (or proteases) act on protein foods and convert them into amino acids, which can be absorbed. *Carbohydrate-splitting* enzymes (or amylases) act on starches or sugars and convert them into simple sugar, such as glucose, which can be absorbed. *Fat-splitting* enzymes (or lipases) act on fats and split them up into fatty acids and glycerin. The fatty acids combine with alkalis found in the intestine to form soap, and this, with glycerin, is absorbed by the cells lining the bowel and built up by them again into droplets of fat.

These chemical changes are the true process of digestion, but the chemical change is greatly assisted by the mechanical breaking up of food into small particles through the mastication of food in the mouth and the churning action of the stomach. This is mere physical change, the changing of a large portion of food into countless minute portions, but it exposes a large surface of the foodstuffs to the action of the enzymes, and therefore hastens the chemical process of digestion.

Chewing and churning break up the proteins and carbo-hydrates, but fat is converted by body warmth into oil, which cannot be chewed. This is coarsely emulsified in the stomach by the churning action and later finely emulsified by the alkalis and soaps present in the small intestine.

Fine division of the food is very important for easy and quick digestion. Hence it is important that food should be slowly eaten and thoroughly chewed. At the same time it is important that food should be enjoyed, as this causes the pouring out of digestive juices containing the enzymes respon-sible for the process of digestion.

18 The Digestive System

THE digestive system consists of:

(1) The *alimentary canal*.
(2) Various *glands* secreting digestive juices which are poured into the canal, where they act upon the food it contains.

The *alimentary canal* is a passage over 9 metres in length, leading through the body from mouth to rectum. It is lined throughout with mucous membrane, which secretes mucus to lubricate the passage of the food through the canal. Its walls are first a coat of loose connective tissue, containing blood vessels and nerves, then a muscular coat, containing longitudinal and circular coats of muscle fibres, and finally an outer coat of fibrous elastic tissue.

The alimentary canal consists of:

(1) The mouth.
(2) The pharynx.
(3) The oesophagus.

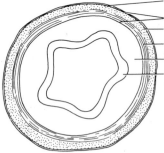

Fibrous elastic (or serous) coat
Muscular coat
Longitudinal muscle
Circular muscle
Loose connective (areolar) coat
Mucous membrane

FIG. 135. Cross-section of the alimentary tract.

(4) The stomach.
(5) The small intestine.
(6) The large intestine, which ends in the rectum.

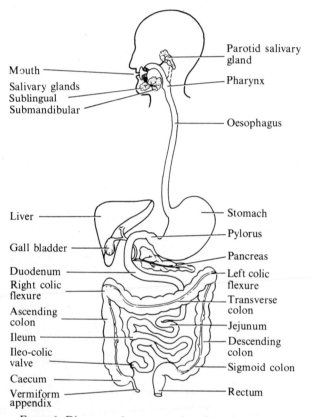

FIG. 136. Diagrammatic representation of the digestive system.

The *glands* secreting digestive juices are:

(1) The six *salivary glands*, which secrete *saliva* into the mouth:

Two parotid, below and in front of the ear.

Two submandibular, under the jaw.
Two sublingual, under the tongue.

(2) The *gastric glands*, which secrete *gastric juice* into the stomach.

(3) The *pancreas*, secreting *pancreatic juice* into the duodenum, the first part of the small intestine.

(4) The *liver*, secreting *bile* into the duodenum.

(5) The *intestinal glands*, secreting *intestinal juice* into the small intestine.

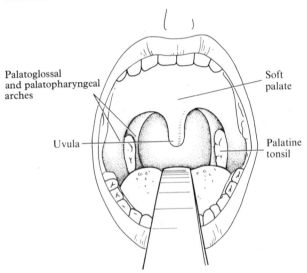

Palatoglossal and palatopharyngeal arches

Soft palate

Uvula

Palatine tonsil

FIG. 137. Looking into the mouth with the tongue depressed.

The Mouth

The mouth is a cavity leading from the external skin to the pharynx. The roof is formed by the hard and soft palates. The soft palate is muscular and carries a fleshy projection, the **uvula**, which hangs down from it in the middle line. The floor is formed by the tongue, which also fills the cavity of the mouth and by mucous membrane, which covers the tongue and runs from it to the gums. The walls are made by the cheek muscles. The opening leading into it is surrounded by the lips, which contain a sphincter muscle (the orbicularis oris), which keeps the mouth closed. The lips are covered with mucous membrane continuous with the outer skin. The opening leading from the mouth into the pharynx or throat is called the fauces. On either side of this opening two folds of mucous membrane can be seen extending down from the soft palate and disappearing below the tongue. These are called the palatoglossal arches and between them lie the palatine tonsils. The mouth is lined throughout with mucous membrane. The alveolar processes of the jawbones project into the mouth and carry the upper and lower sets of teeth. They are covered with mucous membrane to form the gums (see also Fig. 128).

The mouth contains:

    (1) The tongue.
    (2) Teeth.
    (3) Tonsils.

The *tongue* is a muscular organ which occupies the floor of the mouth and rises up into it. It is attached to the inner surface of the mandible at the front and the hyoid bone at the back. It is covered with mucous membrane, and on its upper surface carries numerous little cone-shaped projections called papillae. These increase the surface for the distribution of nerve endings. Underneath the free portion of the tongue is a crescent-shaped fold of mucous membrane running from under its surface to the floor of the mouth called the frenulum.

The *functions of the tongue* are:

    (1) It is the organ of *taste*, being richly supplied with sensory nerves, which end in taste buds in the papillae of the upper surface. Through these we appreciate whether food is sweet, acid, salt, bitter, and also whether it is hot or cold, and hard or soft, through other nerve endings. The flavours of various aromatic substances in food are appreciated by the nerve of smell—e.g. the flavour of onions, bay leaves, raspberries, etc.

    (2) It assists in the *mastication* of food.

    (3) It assists in *swallowing*.

    (4) It assists in *articulation*, moving freely as we speak to alter the shape of the airway and interrupt the passage of air.

The *teeth* consist of:

    (1) A *crown* protruding from the gum.

    (2) A *root* of one or more fangs embedded in the alveolar process of the jaw.

    (3) A *neck* where crown and fang join.

They are made of a very hard substance known as *dentine*, covered in the crown by a still harder material, enamel, which consists of 96 per cent mineral matter, and is the hardest substance in the body. The root is also covered with a hard material, resembling bone, known as cement. In the centre of each tooth is a small cavity, called the pulp cavity, which contains the dental pulp. This pulp consists of blood vessels and nerves to supply the tooth with nourishment and sensation. At the apex of the tooth is a small foramen for the passage of blood vessels and nerves.

There are four types of teeth:

(1) The *incisors*, with chisel-shaped crowns for biting food.
(2) The *canine* teeth with pointed crowns.
(3) The *premolars* or bicuspids, with two cusps or projections for grinding food.
(4) The *molars*, with four or five cusps for grinding food.

There are two sets of teeth:

(1) The temporary teeth.
(2) The permanent teeth.

The *temporary* teeth are twenty in number, ten in each jaw.

	Molars	Canines	Incisors		Canines	Molars
Upper jaw	2	1	2	2	1	2
Lower jaw	2	1	2	2	1	2

The first to be cut are the lower central incisors, which usually come through at about 6 months, though they are in rare cases present at birth. The upper and lateral incisors follow, and all should be cut at about 1 year. The set should be complete between 2 and 2½ years.

The *permanent* teeth are thirty-two in number, sixteen in each jaw.

	Molars	Premolars	Canines	Incisors		Canines	Premolars	Molars
Upper jaw	3	2	1	2	2	1	2	3
Lower jaw	3	2	1	2	2	1	2	3

The first to be cut is the first molar, which lies behind the temporary teeth and comes through at about 6 years. Next, the temporary incisors drop out as their roots are gradually absorbed, and the permanent teeth, which have been developing in the jaw below them throughout infancy and childhood, come through. The permanent set should be complete at 14 years, except for the third molar or wisdom tooth, which does not come through till between 18 and 25 years, or even later. Sometimes it fails to come through, and becomes impacted.

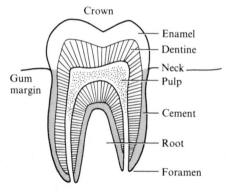

FIG. 138. Section of a tooth, showing its structure.

Normally the grinding teeth in the upper and lower jaws do not lie exactly opposite one another, so that the projections on one set interlock with those on the other for efficient mastication. The lower teeth are half a tooth in front of the upper.

For the development of good teeth it is important that the mother during pregnancy, and the child throughout the growing period, should be liberally supplied with foods rich in calcium, particularly milk and eggs, and with vitamin D, which controls bone formation. It is also important that the jaws are well exercised by breast feeding, followed by the giving of hard foods, so that the jaws and teeth receive a good blood supply. Poor enamel, due to lack of vitamin D or calcium, will result in early decay, as will also the collecting of starchy and sugary foods about the teeth, especially during the night. These foods as they decompose in the mouth produce acids which will act on the calcium of the teeth and make it soluble, causing the teeth to become softer and thus allowing bacteria to enter. Poor jaw development from lack of exercise will cause the teeth to be crowded and out of position, often overlapping; also there may not be room for the third molar tooth (the 'wisdom' tooth) to come through and it becomes impacted.

The *palatine tonsils* are two lymphoid glands lying one on either side of the fauces. They filter bacteria from the lymph returning from the mouth, and deal with bacteria in the food which we eat. They contain deep crypts in which bacterial infection may lodge.

The six *salivary glands* lie around the mouth, and their ducts open into it. The parotid glands discharge saliva by ducts which open opposite the upper molar teeth. The sub-mandibular and sublingual glands discharge saliva into the floor of the mouth.

In the mouth food is:

   (1) Masticated.
   (2) Mixed with saliva.

Saliva is secreted reflexly as a result of pleasant sensations concerning food or the presence of any solid in the mouth. It consists of:

   (1) Water, which moistens the food.
   (2) Salts.
   (3) *Ptyalin*, an enzyme which acts on cooked starch and turns it into malt sugar (maltose).

Mucus is also secreted liberally in the mouth, and lubricates the moving surfaces and coats the food, so that it passes down the oesophagus without damaging the lining membrane.

When the food is well masticated and moistened, the tongue rolls it into a bolus or ball and carries it back into the pharynx to be swallowed.

The Pharynx

The pharynx is a muscular cavity lined with mucous membrane. It is divided into three parts (see Chapter 16). Of these the oro-pharynx and the laryngo-pharynx are associated with swallowing. Food which has been masticated and mixed with the saliva in the mouth is formed into a bolus and pushed into the oro-pharynx by the tongue. The soft palate rises to occlude the naso-pharynx and the larynx rises up under the epiglottis so that the bolus of food can only pass through the laryngo-pharynx into the oesophagus. This is an excellent example of a complicated muscular co-ordination. If the co-ordination is not achieved correctly choking will result.

The Oesophagus

The oesophagus is a collapsible muscular tube 25 to 30 cm long, leading from the pharynx to the stomach. It runs down the neck, behind the trachea, and through the thorax, behind the trachea and heart, finally piercing the diaphragm to reach the cardiac end of the stomach.

The *wall* of the oesophagus consists of:

(1) A lining of mucous membrane, secreting mucus.
(2) Loose connective tissue.
(3) A muscular coat made up of: an outer layer of longitudinal fibres, and an inner layer of circular fibres.
(4) An elastic fibrous outer coat.

The upper one-third is composed of voluntary muscle and the lower two-thirds of involuntary muscle.

The muscular coat has a *peristaltic* action. Peristalsis means a wave of contraction, preceded by a wave of dilatation, passing along the wall and driving the food towards the stomach. It is found throughout the canal and in muscular tubes in other parts of the body.

The Stomach

The stomach is a muscular bag-like organ lying high on the left of the abdomen beneath the diaphragm. It varies in size and shape according to its contents. It is the widest portion of the alimentary canal. Its upper end is pouch-like and is known as the *fundus*; the upper part of the stomach is called the *cardiac end* or portion because it lies under the diaphragm beneath the heart. The oesophagus opens into the cardiac end, the entrance being guarded by a weak sphincter muscle, the cardiac sphincter. This prevents regurgitation of the food into the oesophagus when stooping, unless the stomach contracts very violently as in vomiting. The lower end narrows down like a funnel and curves to the right like the letter J, leading to the duodenum. The exit is guarded by a strong sphincter muscle, the pyloric sphincter; the lower end is therefore known as the *pyloric antrum*. Between the two sphincters are the two *curvatures* of the stomach, the lesser on the upper surface and the greater on the lower surface. The capacity of the normal stomach is about 1 litre, but it can be distended to hold much more.

The *wall* of the stomach consists of:

(1) An outer coat of serous membrane, derived from the peritoneum (for full description of the peritoneum, see p. 238).
(2) A muscular coat.
(3) A layer of connective tissue rich in blood vessels.
(4) A lining of mucous membrane containing the gastric glands.

The serous coat prevents friction.

The muscular coat consists of longitudinal, oblique, and circular fibres, and is especially thick because of the churning action of the stomach.

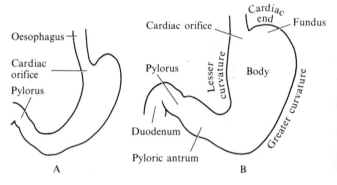

FIG. 139. The stomach. (A) Empty, (B) containing food.

The mucous membrane secretes mucus for lubrication, especially at the cardiac end of the stomach. The membrane is thickly studded with the gastric glands. These are minute tubes, lined with secreting cells, and grouped in pairs, with their ducts opening into the cavity of the stomach. They secrete the gastric juice.

The *functions of the stomach* are:

(1) To churn up the food, further breaking up and softening it.
(2) To digest the food by means of the gastric juice.
(3) The stomach wall also secretes the intrinsic factor

which is necessary for the absorption of vitamin B_{12} (cyanocobalamin). Without this factor normal development of red blood cells does not occur.

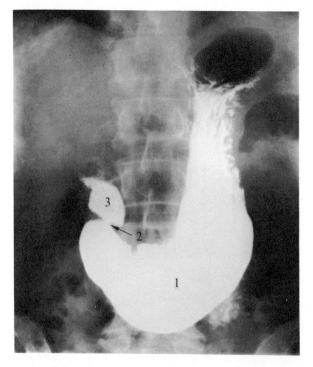

FIG. 140. The stomach. An X-ray of a barium meal, showing the normal stomach, the constriction of the pylorus and the 'duodenal cap', i.e. the beginning of the duodenum, filled with barium. (1) Body of stomach; (2) pylorus; (3) duodenal cap.

The *gastric juice* is secreted by the gastric glands, partly as a reflex following pleasant sensations concerning food, in the same way as the saliva. This causes a copious flow of gastric juice before and during the taking of enjoyable food.

The gastric glands are also stimulated to secrete by a chemical stimulus through an internal secretion or hormone known as *gastrin*. Gastrin is secreted by the lining of the stomach as long as food is present in the stomach. Gastrin is absorbed into the circulation through the stomach wall, and as it circulates in the blood stream it has a stimulating effect on the glands. This hormone therefore maintains secretion after the pleasant sensations connected with eating are over, as long as there is need for it.

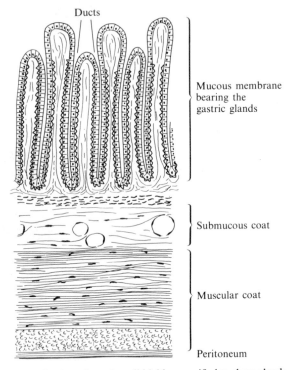

Ducts

Mucous membrane bearing the gastric glands

Submucous coat

Muscular coat

Peritoneum

FIG. 141. Section of gastric wall highly magnified to show glands.

Gastric juice consists of:

    (1) Water, mineral salts and mucus.
    (2) Hydrochloric acid.
    (3) Pepsinogen, which is converted by hydrochloric acid

to the active enzyme pepsin. Pepsin then turns proteins into peptones.

(4) Rennin, which coagulates caseinogen, turning it into casein, and preparing it for the action of pepsin (this is not present in adults).

The juice makes the food more liquid, and acid in reaction. Until it becomes acid, and this may take 15 to 30 minutes in the cardiac end of the stomach, which acts as a reservoir, the ptyalin of the saliva continues to act on the cooked starch. When the food is acid, *pepsin* and *rennin* act on the proteins and on caseinogen. The food is quickly acidified in the pyloric end of the stomach, where peristaltic action is very marked, so that it acts like a mill, churning the food up and mixing it with the gastric juice. The food remains in the stomach for $\frac{1}{2}$ hour to 3 hours or more, according to the nature of the food and the muscularity of the individual stomach. A meal rich in carbohydrate, but containing little protein, as tea, toast, and cake will leave the stomach in $\frac{1}{2}$ hour. A good mixed meal, like a typical dinner, will remain for $2\frac{1}{2}$ to 3 hours or more, though it may leave earlier or stay longer according to the tone and activity of the muscular coat.

The hydrochloric acid in the gastric juice serves several purposes:

(1) It gives the acid reaction required by the gastric enzymes.

(2) It kills bacteria.

(3) It controls the pylorus.

(4) It stops the action of ptyalin.

(5) It converts pepsinogen to pepsin.

The pylorus is normally contracted. When there is food in the stomach the gastric juice makes the contents gradually more and more acid at the pyloric end. When it reaches a certain degree of acidity the pylorus relaxes and a little food passes into the duodenum. The acid food here causes the pylorus to close and the tone of the stomach wall drives food from the cardiac reservoir down to mix with the food in the pyloric end, making it less acid. Gradually the food in the duodenum is made alkaline and that in the pyloric end of the stomach again becomes more acid; this causes the pylorus to open again.

The churning action of the stomach serves to emulsify coarsely any fat which may be present and which the body heat will have melted. This converts the food into a greyish-white fluid called *chyme*.

The Small Intestine

The small intestine is a tube leading from the stomach to the large intestine. It is 6 to 7 metres in length, and lies coiled up in the abdominal cavity. It is divided into:

(1) The *duodenum*, 25 cm long.
(2) The *jejunum* about 2·5 m long.
(3) The *ileum*, about 4 m long.

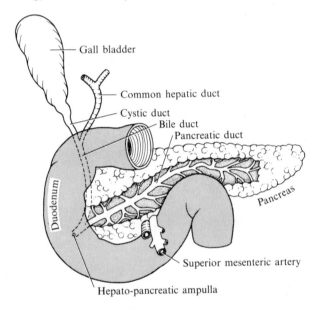

Fig. 142. The duodenum, pancreas and gall bladder.

The *duodenum* lies against the posterior abdominal wall, to which it is fixed by the peritoneum, which covers it. It curves to the left like a horseshoe, and ends behind the stomach. Into it open the bile duct from the liver and the pancreatic duct from the pancreas, by an opening, called the hepato-pancreatic ampulla, which is guarded by a sphincter muscle; this relaxes to allow the digestive juices to flow in, when food enters the duodenum.

The *jejunum* and *ileum* lie coiled up in the front of the abdominal cavity, and are attached to the back wall by a double flare-like fold of peritoneum called the *mesentery*. This carries the blood and nerve supply. The jejunum (from a Latin word meaning empty) is so called because it is always found empty after death. The word ileum means coiled.

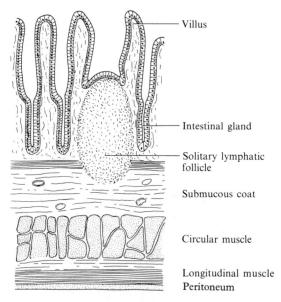

Villus

Intestinal gland

Solitary lymphatic follicle

Submucous coat

Circular muscle

Longitudinal muscle
Peritoneum

Fig. 143. Section of the wall of the small intestine, highly magnified.

The *wall* of the small intestine has the same four coats as the stomach.

(1) A serous coat derived from the peritoneum.
(2) A muscular coat of longitudinal and circular fibres.
(3) A coat of connective tissue rich in blood vessels.
(4) A lining of mucous membrane.

The lining has three peculiarities:

(1) It is *puckered*, forming circular folds which run round the bowel. This increases its extent, so that the lining is

reckoned to be about 30 m in length in the intestine (Fig. 144).

(2) It is covered with fine hair-like projections called *villi*, which give it a velvety appearance and are for the absorption of food. Each villus is covered with cells and contains a network of blood capillaries and a central lacteal or lymphatic capillary (Fig. 145).

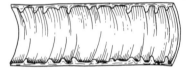

FIG. 144. Section of small intestine, showing puckered lining (circular folds).

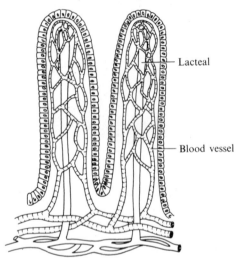

FIG. 145. Villi containing blood vessels and lacteal.

(3) It is supplied with *glands*; these include the *intestinal* glands, which secrete intestinal juice and are little tubular glands, and also glands of lymphoid tissue.

In the jejunum small nodules of lymphoid tissue are present under the mucous membrane, and are known as *solitary lymphatic follicles*. In the ileum the nodules are found in large groups called *aggregated lymphatic follicles*. These are round or oval in outline and can be seen with the naked eye. They may become inflamed in typhoid fever. These lymphoid nodules deal with bacteria which may be absorbed from the contents of the intestine.

The *functions of the small intestine* are:

(1) Digestion.

(2) Absorption of food.

Digestion is carried out by pancreatic juice, bile and intestinal juice.

The pancreatic and intestinal juices are secreted reflexly through sensations concerning food, and also through a hormone called *secretin* produced by the lining of the intestine, as gastrin is secreted by the stomach wall. The juices are alkaline and make the food alkaline in reaction. This finely emulsifies the fat.

The *pancreatic juice* consists of water, alkaline salts and three enzymes acting on the three different foodstuffs:

(1) Trypsinogen, which turns peptones and proteins into amino acids when converted into active trypsin by enterokinase. If active trypsin were secreted by the pancreas, it would be able to digest the protein of the cells which form the gland and its ducts. Trypsin becomes active only when it mixes with the food and the intestinal juice within the bowel.

(2) *Amylase*, which turns starch, cooked and uncooked, into malt sugar (maltose).

(3) *Lipase*, which splits fat into fatty acids and glycerol, after the bile has emulsified the fat, to increase the surface area.

The *bile* contains no enzymes, but is rich in alkaline salts, which serve to emulsify and saponify fats, i.e. make soaps from them.

The *intestinal juice* contains water, salts, and enzymes. These enzymes are:

(1) *Enterokinase*, which converts trypsinogen secreted by the pancreas to active trypsin.

(2) *Peptidase*, which acts on peptones and turns them into amino acids.

(3) *Maltase*, which turns maltose into simple sugar, such as glucose.

(4) *Sucrase*, which turns cane sugar (sucrose) into simple sugar.

(5) *Lactase*, which turns lactose into simple sugar.

(6) *Lipase*, which completes the conversion of fats to fatty acids and glycerol.

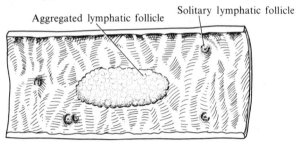

Aggregated lymphatic follicle   Solitary lymphatic follicle

FIG. 146. Lymphatic follicles in the ileum.

These juices are mixed with the food by the muscular action of the wall of the small intestine. In addition to peristalsis, which carries the contents along the tube, contractions occur at intervals along the tube making it look, for the moment, rather like a string of sausages. These movements may be seen on a film taken with the small intestine exposed by an abdominal incision. The contractions occur first at one set of points then at another and are followed by relaxation, having a kneading or churning effect, and bringing the mucous lining into close contact with the contents of the gut.

Absorption. The absorption of proteins, carbohydrates, and fats, takes place almost entirely through the villi in the small intestine. Very little food is absorbed from the stomach, as it is either not yet sufficiently digested, or, if absorbable, e.g. glucose and water, does not stay in the stomach, but merely passes through it. *Proteins* in the form of amino acids and *carbohydrates* in the form of simple sugar are absorbed by the cells covering the villi and pass into the *blood capillaries*, being carried by the portal vein to the liver. Fats in the form of fatty acids and glycerol are absorbed by the cells covering the villi and built up again by them into droplets of fat. These pass into the lymph within the villi and are drained away by the *lymphatic capillaries* or lacteals, which are so named because the lymph they contain is made milk-like by the droplets of fat in suspension. The fat passes via the lymphatic vessels to the cisterna chyli and is carried up the thoracic duct into the blood stream.

The Large Intestine

The *large intestine* is a tube leading from the small intestine to the external skin. It is 1·5 m in length and is divided into: (a) the caecum, (b) the ascending colon, (c) the transverse colon, (e) the descending colon, (e) the sigmoid flexure, and (f) the rectum.

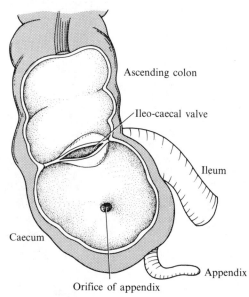

Ascending colon

Ileo-caecal valve

Ileum

Caecum

Appendix

Orifice of appendix

Fig. 147. The caecum, with the appendix.

The *caecum* is the blind end of the colon lying in the right iliac fossa. The ileum enters it from the side, the opening being guarded by a weak sphincter muscle, the *ileo-caecal valve*. It is pouch-like, and to its base the *appendix* is attached. This is a fine, blind tube about 10 cm long. It is known as the vermiform appendix, because it looks rather like a worm. Its wall contains a good deal of lymphoid tissue, hence it is often a site of inflammation.

The *ascending colon* runs up from the caecum on the right side to the undersurface of the liver, where it turns at the right colic flexure to become the *transverse colon*. This runs across the front of the abdomen just below the stomach till it reaches the spleen, where it forms the left colic flexure. This leads to the *descending colon*, which runs down on the left side to the left iliac fossa, where the colon forms an S-shaped bend, the

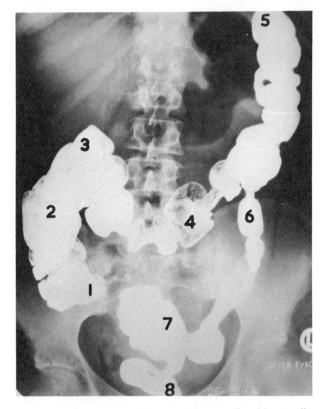

FIG. 148. The colon. (1) Caecum, with appendix; (2) ascending colon; (3) right colic flexure; (4) transverse colon; (5) left colic flexure; (6) descending colon; (7) sigmoid flexure; (8) beginning of the rectum.

pelvic colon or *sigmoid flexure*; this runs into the pelvic cavity and leads to the rectum. The *rectum* is the last 15 cm of the large intestine; it runs straight down the back of the pelvis to join the anal canal; this is about 4 cm long and opens on to the external skin in a circular opening, the *anus*, which is guarded by sphincter muscles (see Figs 148 and 149).

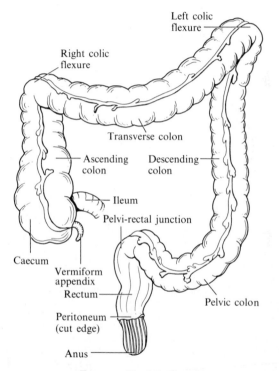

FIG. 149. The large intestine.

The *wall* of the large intestine consists of four coats:

(1) An outer coat of serous membrane—peritoneum.
(2) A muscular coat.
(3) A coat of connective tissue.
(4) A lining of mucous membrane.

The muscular coat is here peculiar. The longitudinal fibres are gathered into three bands called taeniae coli, instead of forming a complete coat. These three bands draw in the wall where they lie, so that it pouches out between them and gives the large intestine a typical puckered or sacculated appearance from the outside. The sacculations are called the haustra coli.

The *functions of the large intestine* are:

(1) The absorption of water and salts.
(2) The excretion of faeces.

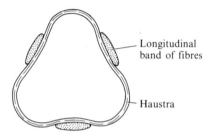

Longitudinal band of fibres

Haustra

FIG. 150. Cross-section of the large intestine.

The material which enters the large intestine consists of water, salts, very little food material, as this has been digested and absorbed in the small intestine, cellulose, which is indigestible, and bacteria. The bacteria are very numerous, as, though they are largely killed in the stomach, the alkaline reaction, food, warmth and moisture in the small intestine encourages their growth here. The material is in a very fluid state. In the colon water and salts are quickly absorbed, so that the fluid is rapidly turned into a paste containing the cellulose and bacteria, many of which die from lack of water and food. This paste forms the faeces, which consists of a fluid and solid part, the latter being roughly 50 per cent cellulose and 50 per cent dead bacteria. At intervals mass movements propel the faeces into the rectum, from which it is excreted.

Defaecation, or the passing of the faeces. The rectum is normally empty. When faeces enter it a sensation of fullness is produced. In the infant and lower animals this sets up a reflex relaxation of the sphincters, which is followed by the emptying of the rectum through peristaltic action in the bowel. In man the act is under voluntary control. The sensation of fullness produces the desire to empty the bowel, but the act can be controlled till a convenient opportunity occurs. It is, however, essential that the opportunity should be found when the sensation occurs, otherwise it will pass off and the bowel cannot then be emptied and constipation occurs. The desire is likely to occur after the taking of food or drink, particularly the drinking of cold water, due to a reflex called the gastro-colic reflex. It is important to train children into a regular habit of emptying the bowel each morning either after break-fast or after a drink of water. The longer the residue is retained, the more water is absorbed from it and the more difficult it becomes to pass.

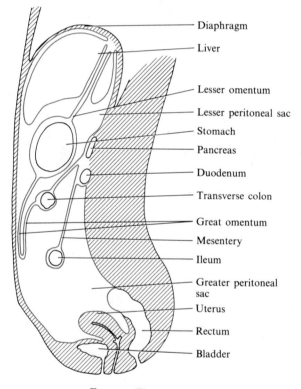

FIG. 151. The peritoneum.

The Peritoneum

The peritoneum is a **serous membrane** lining the abdomen and covering the abdominal organs. It is similar to the pleura but more complicated as it covers a number of organs.

The functions of the peritoneum are:

(1) To prevent *friction* as the abdominal organs move on one another and against the abdominal wall, as its free surfaces are always moist with serum and are smooth and glistening.

(2) To *attach* the abdominal organs to the abdominal wall, except in the case of the kidneys, duodenum and pancreas, which lie behind it. The ascending and descending colon are covered only on their anterior surface by peritoneum. This

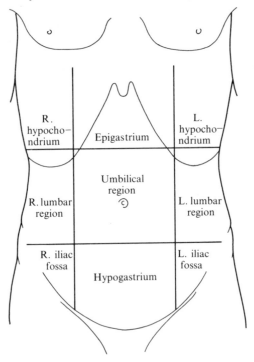

FIG. 152. Regions of the abdomen.

means that only the transverse or sigmoid colon can be brought on to the anterior abdominal wall for a colostomy operation.

(3) To carry *blood vessels*, lymphatics and nerves to the organs; these run between the two folds of the membrane.

(4) To *fight* infection, as it contains many lymphatic nodes.

Several portions have special names, and of these the most important are:

(1) The great *omentum*, a double fold of peritoneum which hangs down from the lower border of the stomach and loops up again to the transverse colon; this carries a great deal of fat. It is frequently found adhering to inflamed surfaces; to prevent the spread of infection.

(2) The *lesser omentum*, a fold of peritoneum which extends from the inferior surface of the liver to the lesser curvature of the stomach.

(3) The *mesentery*, which attaches the small intestine to the back wall of the abdomen and is rather like a flare, having a very long border along which the bowel runs, but arising from a very short line at the back of the abdomen.

The layer lining the abdominal wall is called the parietal peritoneum, and that covering the organs the visceral peritoneum.

Regions of the Abdomen

The abdomen is divided into nine regions for purposes of description by two transverse and two upright lines (see Fig. 152). In describing the position of organs, the regions in which they lie may be used, e.g. the stomach lies in the left hypochondriac, epigastric, and umbilical regions. The kidneys are in the right and left lumbar regions respectively. The caecum is in the right iliac fossa. The bladder rises when full into the hypogastric region.

19 The Liver, Biliary System and Pancreas

The Liver

THE liver is the largest gland in the body. It weighs about 1·4 kg, and is a dark reddish-brown colour. It lies immediately beneath the diaphragm on the right side of the abdomen. On the under surface is a small pear-shaped bag, the *gall-bladder*. The liver is divided into two main *lobes*, the right and left lobes. The right lobe is the larger and lies over the right colic flexure of the colon and the right kidney. The left lobe is comparatively small and lies over the stomach. The lobes are

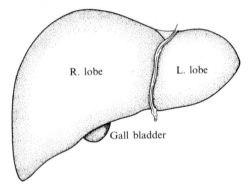

R. lobe   L. lobe

Gall bladder

FIG. 153   The liver, general appearance.

made up of small *lobules*; these are composed of the typical *liver cells*, which are large nucleated cells with many important functions.

Into the liver pass the portal vein and the hepatic artery, bringing blood rich in foodstuffs and oxygenated blood respectively. From the liver run the three hepatic veins carrying venous blood to the inferior vena cava, and the right and left hepatic ducts carrying bile. The common hepatic duct joins the cystic duct from the gall bladder to form the bile duct, which enters the duodenum (see Fig. 157).

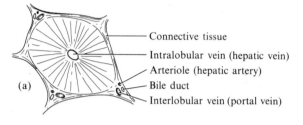

(a)
- Connective tissue
- Intralobular vein (hepatic vein)
- Arteriole (hepatic artery)
- Bile duct
- Interlobular vein (portal vein)

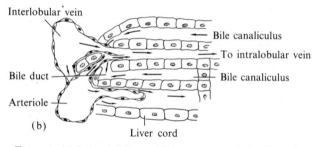

Interlobular vein

Bile duct

Arteriole

(b)

Liver cord

Bile canaliculus

To intralobular vein

Bile canaliculus

FIG. 154. (a) A liver lobule; and (b) the scheme of blood vessels and bile ducts.

Within the liver the portal vein and hepatic artery divide up into small vessels which run between the lobules of the liver (interlobular veins and arteries). These provide a capillary network within the lobules among the actual liver cells. These capillaries drain into a vein in the centre of each lobule, which in turn drains into veins which empty into the hepatic veins. These carry away both the blood from the portal capillaries and the oxygenated blood brought to the liver by the hepatic arteries.

Between the lobules there are also small bile ducts into which fine bile capillaries empty the bile secreted in the lobules; the bile ducts empty into the right and left hepatic

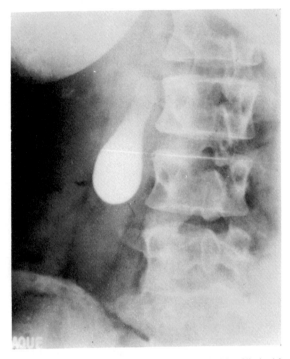

Fig. 155. An X-ray showing the normal gall-bladder filled with an opaque drug (e.g. Opacol).

ducts, which join to form the common hepatic duct. This in turn joins the cystic duct to form the bile duct. If the bile is not required for the purposes of digestion it passes up the cystic duct into the gall bladder, where it is both stored and concentrated. The *gall-bladder* can only hold from 30 to 60 ml, but its wall is able to absorb water, so that the bile it contains becomes gradually more and more concentrated. When food enters the duodenum the sphincter at the mouth of the bile duct relaxes, and the bile stored in the gall-bladder is driven into the intestine by the contraction of the wall of the gall-bladder.

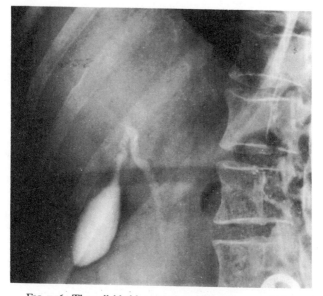

FIG. 156. The gall-bladder, emptying after a fatty meal, the opaque media showing the bile ducts.

The functions of the liver can be divided into three sections:

Metabolic Functions

(1) Stored fat is broken down to provide energy. This process is called desaturation.

(2) Excess amino acids are broken down and converted to urea.

(3) Drugs and poisons are de-toxicated.

(4) Vitamin A is synthesized from carotene.

(5) The liver is the main heat-producing organ of the body.

(6) The plasma proteins are synthesized.

(7) Worn out tissue cells are broken down to form uric acid and urea.

(8) Excess carbohydrate is converted to fat for storage in the fat depots.

(9) Prothrombin and fibrinogen are synthesized from amino acids.

(10) Antibodies and antitoxins are manufactured.

(11) Heparin is manufactured.

Storage Functions

(1) Vitamins A and D.

(2) Anti-anaemic factor.

(3) Iron from the diet and from worn out red blood cells.

(4) Glucose is stored as glycogen and converted back to glucose in the presence of insulin as required.

Secretory Functions

Bile is formed from constituents brought by the blood.

The storing of glucose. All the glucose absorbed by the villi of the small intestine is brought by the portal vein to the liver; the liver cells convert the excess glucose not required for the immediate needs of the tissues into insoluble glycogen, allowing only what is needed to pass into the blood stream. This excess is stored as glycogen in the liver cells until required for use, when it is reconverted into glucose, which the blood carries away in solution. This activity is controlled by the insulin secreted by the pancreas. If there is a lack of insulin, glucose cannot be turned into glycogen in the normal way and there is too much sugar in the blood, a condition called hyperglycaemia.

The formation of urea. The amino acids derived by the process of digestion from the protein food we eat are absorbed by the villi of the small intestine and brought by the portal vein to the liver. The amino acids required to make good the wear and tear of tissue and produce its growth are allowed to pass straight through the liver into the blood stream. Others are used to form the blood proteins. Any excess protein or second-class protein which is unsuitable for tissue-building is broken up in the liver to form: (a) body fuel, composed of carbon, hydrogen and oxygen; and (b) urea, a compound containing the nitrogen present in all proteins, which is incombustible, and therefore useless unless needed for the building of tissues. This urea is a soluble substance which the blood stream carries from the liver to the kidneys for excretion from the body.

The secretion of bile. The bile is a thick greenish-yellow fluid secreted by the liver cells. It is alkaline in reaction. The liver secretes on an average about a litre of it per day. The bile consists of water, bile salts and bile pigments. The bile salts give the bile its alkaline reaction and include both organic and inorganic salts. Among the former is the substance cholesterol, which is the main ingredient of the typical gall-stone. The bile pigments are derived from the haemoglobin of the worn-out red corpuscles which are excreted from the body by this channel, giving the normal colour to faeces: they are also present in the blood and colour the urine.

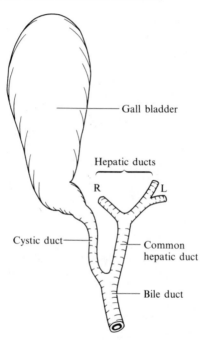

FIG. 157. The gall-bladder and its ducts.

The *uses of bile* are:

(1) It helps to *emulsify* and *saponify* fats in the small intestine by its alkalis. In this way the surface area is increased and the action of enzymes is increased.

(2) It stimulates *peristalsis* in the intestine, so that it acts as a natural aperient.

(3) It is a channel for *excretion* of pigments and toxic substances from the blood stream, such as alcohol and other drugs.

(4) It acts as a *deodorant* to the faeces, lessening their offensive odour. It is suggested that this may be due merely to the fact that a lack of bile means poor digestion of fat, so that fat remains in the intestine in excess, coating the other foods and preventing their digestion and absorption. As a result undigested protein is attacked by bacteria and, decomposing, produces an excess of sulphuretted hydrogen, the gas which causes the smell of abnormal faeces, foul drains and rotten eggs.

From the list of its functions the importance of the liver is very obvious and it is essential to life. On the other hand, it is a large organ able to undertake more work than is normally demanded of it, and seven-tenths of the organ may be destroyed by disease before death occurs from hepatic inefficiency.

The Pancreas

The pancreas is a compound saccular or racemose gland lying across the back of the abdomen behind the stomach. It is shaped roughly like a dog's tongue, with the tip extending to the spleen and the rounded end fitting into the curve of the duodenum. It is 12 to 15 cm long. It produces both an external and an internal secretion. The external secretion is the *pancreatic juice*. This is secreted by the countless little saccules of the gland and collected into a main duct which runs the length of the gland and empties into the duodenum together with the bile duct. The pancreatic juice plays an important part in the process of digestion in the small intestine (see p. 231 and Fig. 143).

The internal secretion is *insulin*. This is produced by little groups of cells lying between the saccules of the gland that secrete the pancreatic juice. These small groups of cells are richly supplied with blood capillaries and are called the interalveolar cell islets (islets of Langerhans). They form only a small part of the gland, and yet their secretion is essential to normal life. Insulin controls the metabolism of carbohydrates. Its presence in the blood is essential to allow the liver and muscles to store and burn glucose. If there is a lack of it the condition of diabetes mellitus will occur. The individual is unable to store sugar in the normal way, so that there is excess

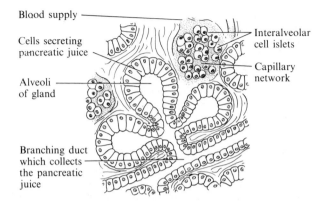

Blood supply

Cells secreting
pancreatic juice

Alveoli
of gland

Branching duct
which collects
the pancreatic
juice

Interalveolar
cell islets

Capillary
network

FIG. 158. The minute structure of the pancreas, showing the saccules secreting the pancreatic juice and the islet cells which secrete insulin.

glucose in the blood, but the tissues cannot burn it and it is excreted in the urine. In the meantime the body must have fuel and burns fat in place of glucose. Since fat will not burn completely unless sufficient glucose is burnt with it, fat combustion is to some extent incomplete and results in the formation of acetone bodies. In severe cases these are produced in large volume and give rise to diabetic coma, which will prove fatal unless the patient is given insulin to enable him to burn more sugar and so break down fats completely.

20 Metabolism—Water and Electrolyte Balance

Metabolism

Metabolism is the term applied to all the changes that occur in the body in connection with the use of food. The Greek word *metabolé*, meaning change, provides the derivation of the word 'metabolism', which covers all the changes concerned with the use of food by the body tissues, the formation of waste products from its use, and the excretion of those waste products.

The changes included in the process of metabolism are of two distinct varieties:

(1) Building-up changes, which are called *anabolic changes* or *anabolism*, e.g. the building up of muscle from the amino acids obtained from proteins, or of fat from fatty acid and glycerol.

(2) Breaking-down changes, which are called *catabolic changes* or *catabolism*, e.g. the breaking down of glucose or fat into carbon dioxide and water to release energy for activity.

This division is now regarded as a little artificial as these metabolic changes are due to enzyme action and within the body cells they take place continuously and simultaneously; for example, a muscle cell constantly builds up fresh protoplasm from the amino acids provided by the blood stream; at the same time it requires energy from the burning of fuel, such as glucose, for this activity and the activity of contraction, and breaks up worn-out protoplasm into waste products, such as urea, for excretion.

Metabolic changes take place all the time in every living thing, but are increased during activity, such as movement or the digestion of food, and are decreased during rest. *Basal metabolism* is the metabolism that goes on when the body is at absolute rest. The *basal metabolic rate* (BMR) is sometimes calculated as an indication of the presence or absence of disease, since over-activity of the thyroid gland raises the rate and under-activity lowers it.

The basal metabolic rate is calculated by measuring the amount of heat produced in the body, either directly in a respiratory chamber, or indirectly, by measuring the amount of carbon dioxide produced, and calculating from it the quantity of oxygen used. The test is made after rest in bed for the night, at least 12 hr after taking a meal, with the individual relaxed and in as tranquil a state of mind as possible. The indirect method is generally used, and in this the expired air is collected in a bag, over a period of time, usually five minutes; from this the amount of carbon dioxide expired is measured, and the amount of oxygen used and therefore the amount of heat produced in the body is calculated. The normal basal metabolic rate is expressed as a percentage of the normal rate; thus a BMR of + 10 means that the rate is 10 per cent above the normal for the individual, taking into consideration age, weight, height, and sex; a BMR of − 10 means that it is 10 per cent below normal. The rate is normally higher in young persons, in males and in persons with a large surface area and therefore a greater heat loss. Metabolic rate is calculated per square metre of the body surface area. The surface area is assessed from a person's height and weight: 1·8 metres is an average surface area for an adult male, and 40 calories per metre per hour is an average metabolic rate.

The metabolism of each foodstuff must be considered separately.

Carbohydrate Metabolism

Carbohydrates in the form of starch and sugar are acted on by the enzymes of the saliva, pancreatic and intestinal juices, and converted into simple sugars such as glucose ($C_6H_{12}O_6$). This simple sugar is absorbed by the villi of the small intestine, passes into the blood capillaries, and is carried by the portal vein to the liver, where excess sugar is stored, as glycogen.

The use of glucose is to serve as the main *body fuel* for the production of energy for work and heat. The glucose required for immediate use is carried straight through the liver into the hepatic veins and inferior vena cava, and so enters the circulation. Any excess not required for the body's immediate needs is converted by the liver cells into *glycogen*, which is insoluble and is stored in the liver until required for use. Glycogen in the animal world is similar to starch in the plant world. Both are insoluble and are formed from sugar by the splitting off of water from it. When sugar is needed by the body the glycogen is converted back into glucose, which the body fluids dissolve, so that it passes into the blood stream. Both the formation of glycogen from glucose and the forming of glucose from glycogen are the work of enzymes produced by the liver cells.

Glucose is specially required by the most active tissues of the body, the muscles and glands, but all tissues need it to some extent. The muscles, like the liver, are able to store it to a slight extent in the form of glycogen. The tissues of the body utilize their fuel very economically. Glucose is burnt to produce energy for muscle contraction, but more energy is released by complete combustion than is required for the work of the muscles, and the excess energy is used to build up glycogen again from some of the partially burnt fuel. Actually it is estimated only about one-fifth of the fuel is completely burnt down into the waste products carbon dioxide and water. The remaining four-fifths, after partial combustion, is built up again into glycogen ready for use when the muscles require it. The energy for this building-up work is obtained from the complete combustion of the other fifth of the fuel.

The *waste products* of the combustion of the carbohydrates are *carbon dioxide* and *water*, which are carried away by the blood stream and excreted from the body. The lungs excrete the carbon dioxide and water, which is also got rid of by the skin and kidneys. Combustion of glucose may be incomplete if there is deficient oxygen supply, and gives rise to acid bodies which cause the pain of acute cramp, as occurs in violent exercise in the muscles or heart wall. The acids cause the blood vessels to dilate, which increases the blood supply, and thus the oxygen supply, and results in the passing off of the pain, as combustion can then be completed.

The metabolism of carbohydrate is controlled by *insulin*, the internal secretion of the pancreas (see p. 246). Without insulin the tissues are not able to burn glucose nor the liver to store it as glycogen. If the insulin supply is normal the amount of glucose in the blood varies very slightly and is normally about 80 to 120 mg per 100 ml of blood. After a meal it will rise slightly, the rise immediately stimulates the pancreas to make more insulin, and the extra insulin causes the liver and muscles to store any excess quickly and the blood sugar falls to normal again. If there is deficiency of insulin the blood sugar rises too high, and neither liver nor muscles can store glucose to the normal extent. This occurs in diabetes, a condition in which the pancreas is diseased and fails to produce the normal quantity of insulin, so that the blood contains excess sugar, and sugar is excreted through the kidney into the urine. Fat is burnt instead of sugar and this is dangerous (see p. 247). On the other hand, too much insulin in the blood causes excess glucose to be stored and leaves insufficient in the blood stream. This is very serious and may even cause convulsions, coma, and death. It is chiefly seen in the giving of an overdose of insulin.

Protein Metabolism

Proteins are converted by the enzymes of the gastric, pancreatic, and intestinal juices into amino acids. These are absorbed by the villi of the small intestine and carried by the portal vein to the liver.

The chief use of protein is to provide *material for body building*. The new tissue required for the purposes of growth and repair can only be made from this foodstuff, since no other foodstuff contains the nitrogen essential for the making of a living cell. The amino acids required for tissue building pass through the liver and are carried by the blood stream to all parts of the body for this purpose.

Protein can also be used as *body fuel*. Excess protein and protein unsuitable for body building, such as the second-class protein obtained from plant foods, are split up in the liver to form:

(1) *Body fuel* in the form of glucose (containing carbon, hydrogen, and oxygen).

(2) *Urea* or nitrogenous waste matter (containing nitrogen, which is incombustible, and hydrogen).

This process is termed the deamination of the amino acids. The nitrogenous content of the amino acids not required for body building is converted first into ammonia which, for the most part, combines with carbonic acid and splits up into urea and water in the liver.

The glucose is either burnt or stored as required. The urea is readily soluble, and, being useless for fuel, is carried away by the blood stream and excreted from the blood by the kidneys.

Protein will also serve as fuel in conditions of starvation. Combustion must go on continuously to maintain life. If there is lack of other fuel, protein will be sacrificed for use as fuel, even though it is needed for processes of repair and growth. As a result, weakness and loss of weight will follow when either the body cannot obtain food, as in famine, or cannot digest and absorb it, as in disease.

The *waste products* of protein metabolism are *urea* and to a lesser extent *uric acid* and *creatinine*. Uric acid is less soluble than urea, and comes particularly from the nuclear material in our food. Creatinine is the waste product of the breaking down of our own body protein. All these protein wastes are excreted by the kidneys in the urine. About 30 g of urea leave the body each day in this way, as also do traces of uric acid and creatinine.

Fat Metabolism

Fat is emulsified by alkalis and split up into fatty acids and glycerol by the lipase of the pancreatic juice. These substances are absorbed by the lacteals and pass through the lymphatic circulation up the thoracic duct into the blood stream.

The use of fat is to serve as *fuel* for the production of heat and energy in the tissues. Fat is a better fuel than glucose in that it produces twice as much heat and energy per gramme of fuel used. On the other hand, it is less easy to digest and absorb and less satisfactory to burn. Provided it is burnt with sufficient sugar it is completely burnt and converted into carbon dioxide and water. If there is little or no sugar to burn with it, combustion is incomplete and ketone bodies are formed (acetone and diacetic acid), which in small quantity produce a sense of fatigue, and in large quantity alter the reaction of the blood, reducing its alkalinity and causing the condition known as acidosis, which results in drowsiness, coma, and finally death, if the condition is not corrected by providing the tissues with glucose. This acidosis occurs in starvation and in diabetes, and in the second condition insulin must be provided as well as glucose to enable the tissues to make use of this fuel.

Fats, before they can be used as body fuel, must be prepared for combustion in the tissues by the liver. This is again a chemical process carried out by the liver cells, and is known as the desaturation of fats.

Fat is also required for the building of various tissues, e.g. nerve tissue, fatty tissue and marrow; fat derivatives are found in the secretions of certain glands.

Fat not required for immediate use as fuel can be stored as fatty tissue. This is found particularly in the subcutaneous tissue and in the body cavities, but whereas sugar can only be stored in very limited quantities, the total being about 225 g in liver and muscles, fat can be stored in very large quantities, and may be found in large amounts even in the muscles themselves. This is, however, undesirable, as it causes increase in weight, which limits activity, setting up a vicious circle.

The putting on of fat does not necessarily mean that too

much fat is being eaten, as the body can convert excess glucose into fat for storage, and excess protein not required for body building into glucose. The metabolism of the three foodstuffs is therefore closely linked up, and the taking of food of any kind in excess of the needs of the tissues will lead to an increase in weight.

The *waste products* of fat metabolism are *carbon dioxide* and *water*, if combustion is complete; these are excreted by the lungs, skin, and kidneys, as in the case of carbohydrates. If combustion is incomplete, the *acetone bodies* formed also leave the body by the same routes. The volatile acetone can be smelt in the 'sweet' breath, and acetone and diacetic acid can be found in the urine.

Water and Electrolyte Balance

Water is one of the essential foodstuffs, but it is a simple substance which can be absorbed and used in the body without chemical change. It enters into many of the metabolic changes which occur in the body, combining with proteins, carbohydrates and fats in digestion and being split off from them when they are used as fuel to produce energy. It is general today to consider the balance rather than the metabolism of water and salts, and the electrolytes which the salts form in the body.

Water forms the greater part of the body cells and the body fluids. Roughly two-thirds or more accurately 60 per cent, of the body weight consists of water. This proportion must be maintained since all the complicated processes of life demand its presence. Of this water 70 per cent is inside the body cells (*intracellular*) and the remaining 30 per cent is in the body fluids (*extracellular*); 15 to 20 per cent being in the interstitial spaces in the tissues, bathing all the body cells, even the bone cells, and the remaining 10 to 15 per cent forming the fluid of the blood, i.e. the plasma and the lymph. These three volumes of fluid are separated only by thin semi-permeable membranes, the cell walls and the capillary walls; water constantly passes through these walls from one of these areas to the others, though the volume of each remains remarkably constant in normal health.

The water of which our bodies are largely made is, however, not static. Fresh water is taken in by the body each day and passes out of it by a number of channels. The total quantities taken in and passed out must balance one another. Water is taken in as water and other fluids drunk, and also in the foods eaten, which, like the body, consist largely of water. On an average the healthy man takes in 1·5 litres of fluid as water and other drinks daily, and a little over 1 litre in his food, an average of 2600 to 2800 ml. A similar quantity is passed out of the body by the lungs as water vapour (400 to 500 ml), by the skin as sweat (500 to 600 ml), by the kidneys as urine (1000 to 1500 ml) and in the faeces a small quantity (100 to 150 ml).

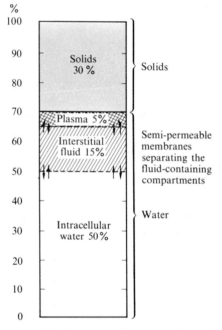

FIG. 159. Diagram to show the average percentage proportions by weight of solids and water, and the distribution of water in the body. The arrows indicate the passage of water and solutes between the three compartments containing them.

The quantities lost in urine, sweat, and water vapour from the lungs vary with conditions. In hot weather and heavy work more sweat is produced to cool the body, and less urine is passed; at the same time the loss of fluid causes thirst, so that more fluid is drunk. In fever the same changes take place; there is increased loss of fluid and there must be increased intake to balance it.

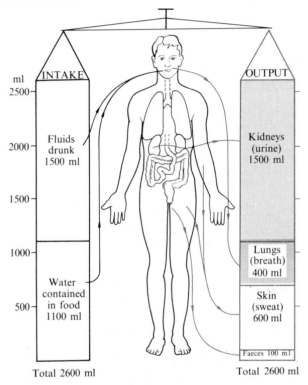

FIG. 160. Diagram to show how water balance is maintained by various organs.

In addition to the balance of the quantity of water in the body, the body fluids must be of the right reaction, that is they must contain the correct balance of electrolytes. These electrolytes are minute particles split off from the molecules of the various salts and dissolved in water. They are called ions (see p. 8). They carry electric charges and are of two types, negatively-charged particles (anions) and positively-charged particles (cations). The total of negatively-charged ions must balance the total of positively-charged ions. The chief negatively-charged ions are chloride (Cl), bicarbonate (HCO_3), and phosphate (PO_4); chloride and bicarbonate are present in quantity in plasma and interstitial fluid, while intracellular fluid contains mainly phosphate. The chief positively-charged ions are sodium (Na) and potassium (K), with a little calcium (Ca) and magnesium (Mg). Sodium is present as the main positive ion in plasma and interstitial fluid, while intracellular fluid contains chiefly potassium.

Salts and the ions produced from them are constantly taken into and lost from the blood. Salts are contained in all foodstuffs, but added salt, used in cooking and in eating, is necessary to maintain the normal electrolyte balance and replace the salt lost in urine and sweat; 3 to 4 g of sodium chloride, or common salt, is needed daily.

The quantities of the various salts are estimated rather as milli-Equivalent/litres (mEq/l) of the ions they produce (see p. 11). There are normally in plasma 155 mEq/l of negatively-charged ions, balanced by 155 mEq/l of positively-charged ions. The bulk of these are provided by chloride (102 mEq/l) and sodium (143 mEq/l). These figures are given for reference, as the nurse will see them in laboratory reports to check for deficiencies or excess of sodium, potassium, chloride, bicarbonate, etc., in the blood. These may result from deficiency in intake or excessive or reduced loss, especially in abnormal conditions affecting the kidney, or from heavy sweating or fever.

Acid–Base Balance

The acid–base balance of the body fluids affects their reaction. Normally the body fluids are slightly alkaline and vary very little during life, since this reaction must be correct for the action of the various enzymes which control the cell activities, such as digestion, the use of food for growth and energy production, and the production and excretion of waste products. Normally venous blood is a little less alkaline than arterial blood because of the carbon dioxide and other acids produced in the tissues. The interstitial fluid and the intracellular fluid are similar but slightly more alkaline. This balance is maintained by the 'buffers' these fluids contain. Alkalis and proteins neutralize acids produced by tissue activities, preventing acidosis, or ketosis, a condition which leads to coma and death if the body fluids become less alkaline than normal. (The normal average is pH 7·4; see Chapter 1.) In the same way, acids such as carbonate and chloride neutralize excess alkalis in the body fluids, preventing alkalosis, a condition which also can prove fatal if it is not corrected. It may be caused by taking in an excess of alkaline salts such as sodium bicarbonate, or by loss of acid from excessive vomiting or from continuous gastric aspiration after operation, or in respiratory conditions, or from retention of abnormal quantities of an alkali such as potassium, through renal failure.

21 Endocrine Glands

THE ENDOCRINE glands are a group of glands which produce secretions that they pour directly into the blood stream. These secretions are termed hormones. They are made by the cells of the gland from materials picked out from the blood vessels with which the ductless glands are well supplied. Being made from the blood, they are composed of substances obtained from food; they are largely protein derivatives, but some are derived from fat and are sterols similar to cholesterol, which is present in bile.

The ductless glands are widely scattered in the body and include:

(1) The *pituitary* gland or hypophysis.
(2) The *adrenal* glands.
(3) The *thyroid* gland.
(4) The *parathyroid* glands.
(5) Parts of the *pancreas*, which was described in Chapter 19.
(6) Parts of the *ovary* and *testes*, which will be described in Chapter 27.

Two other glands, the *thymus* and the *pineal body*, are believed to belong to this group of ductless glands, though their functions are not known.

Although these glands have been described and investigated as separate organs, there is little doubt that functionally they are closely interrelated. When all the secretions from these ductless glands are present in their correct proportions, they form, together with the other fluid constituents of the blood, the correct medium to bathe the cells of the tissues, which, in consequence, are healthy and grow normally. When one constituent is deficient or excessive, the balance is upset and the cells suffer.

Tissue activity in the body is produced in two distinct ways:

(1) *By nervous stimulation*, through the nerve supply to the tissues;

(2) *By chemical stimulation*, through the hormones or internal secretions in the blood and tissue fluid.

Roughly, nervous stimulation is used to produce rapid and specialized activities, such as muscular actions or processes of thought, which are completed in seconds or minutes. Chemical stimulation through hormones produces general effects such as growth, sexual development and the metabolism of foodstuffs—processes which can be measured in hours, months or years. Exceptions to this are the hormones produced by the medulla of the adrenal glands and the posterior lobe of the pituitary gland, both of which are intimately linked with the autonomic nervous system.

Knowledge of the ductless glands has for the most part been acquired during the last 30 to 40 years. It makes an interesting chapter in medical history, but the knowledge is still incomplete and is constantly being increased. The first breakthrough was the discovery of insulin by Dr Banting and Dr Best in Canada in 1922, in spite of great difficulty due to lack of financial and medical support.

In the first place the functions of these glands were discovered by noting the symptoms which occurred when the glands were affected by disease, either overgrowth or destruction of the affected organ. Attempts were made to obtain extracts from the glands, but these were ineffective at first as the hormones were destroyed in the process. Grafts of whole glands or sections of gland tissue were introduced into the body. Actually the grafted tissue died but some secretion was present in it and was absorbed and had some effect.

With advances in scientific knowledge and improved equipment and laboratory facilities over recent years, gradually one hormone after another has been isolated, obtained in pure form, analysed and in some cases successfully synthesized; for example thyroxine, adrenaline and cortisone. This is interesting because the enzymes, produced by the glands with ducts, have been known over a longer period of time but have not yet been made in the laboratory.

The endocrine glands secrete their hormones continuously, but the quantity of secretion can be increased and decreased according to the body needs. Secretion is largely controlled by chemical stimulation through hormones of other glands; for example, the pituitary gland produces hormones which stimulate the thyroid, adrenal cortex and the reproductive glands. As the quantity of the hormones of these glands in the blood increases the secretion of the pituitary hormones is decreased.

This is called a 'feedback mechanism'. The medulla of the adrenal glands is an exception, being controlled by the sympathetic nerve supply and increasing the output of its hormones in response to emotion and stress such as fright, anger and anxiety.

The Pituitary Gland or Hypophysis

The pituitary gland is a small gland about the size of a pea. It lies in the pituitary fossa (sella turcica) in the base of the skull, and is attached by the neural stalk to the hypothalamus at the base of the brain. It is sometimes regarded as the master gland of the body, since it exerts, through its hormones, an influence on nearly all the other ductless glands. However, the glands

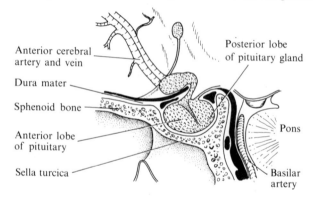

Anterior cerebral artery and vein

Dura mater

Sphenoid bone

Anterior lobe of pituitary

Sella turcica

Posterior lobe of pituitary gland

Pons

Basilar artery

FIG. 161. The pituitary gland, shown in section, lying in the sella turcica or pituitary fossa. Note the stalk joining it to the base of the brain

really act in unison, the pituitary becoming active if the others start to fail and reducing its output when their hormones increase in amount in the blood stream. Although so small, it consists of an anterior and a posterior lobe. These two lobes, in spite of the fact that they form one organ, are different in structure, origin and function. The anterior lobe arises from epithelial tissue, derived in the early stages of development from the lining of the alimentary canal, while the posterior lobe is a down-growth from the base of the brain and remains connected to it by a small stalk. This direct connection with the brain accounts for the influence which the hypothalamus, a nerve centre in the base of the brain, is thought to exert over the gland. The hypothalamus controls many metabolic processes in the body, largely through its link with the posterior lobe of the pituitary gland.

The *anterior lobe* is the larger and produces a number of hormones of great importance, namely:

(1) The *growth hormone*, affecting particularly the growth of bones and muscles and so determining the size of the individual. Over-secretion of this hormone in children causes gigantism. In adults it causes overgrowth of the head, particularly affecting the lower jaw, the hands, and feet; this condition is known as *acromegaly*, which means in Greek 'large extremities.' Undersecretion in children produces dwarfism. These dwarfs and giants produced by under- or over-activity of the anterior lobe of the pituitary gland are normally proportioned and usually of normal intelligence, unlike the cretin or hypothyroid dwarf.

(2) The *thyrotrophic** hormone, which stimulates the thyroid gland.

(3) The *adrenocorticotrophic hormone* (sometimes referred to as ACTH), which stimulates the adrenal cortex.

(4) The follicle stimulating hormone (FSH) which affects the development of the ovarian follicles in the female and stimulates the production of sperm cells in the testes of the male.

(5) The luteinizing hormone (LH) in the female which acts on the corpus luteum. In the male this hormone is called the interstitial cell stimulating hormone and acts on the testes.

(6) The lactogenic hormone is active only after the delivery of a baby. It stimulates the secretory cells in the breasts to produce milk.

The *posterior lobe* produces two hormones: (a) Oxytocin, which causes the muscles of the pregnant uterus only, to contract. It also causes the ducts of the mammary glands to contract, helping to express the milk which the gland has secreted into the ducts. (b) The antidiuretic hormone has a direct effect on the tubules of the kidney, increasing the amount of fluid they absorb so that less urine is excreted.

It is suggested that these two substances may be produced in the hypothalamus itself and merely pass down the stalk of the pituitary gland to be stored in the posterior lobe of the pituitary gland and liberated from it into the circulation.

* The word *trophic* comes from a Greek word meaning nourishment.

Damage to the posterior pituitary lobe, its stalk, or its connections with the brain, produce diabetes insipidus, in which there is an increased output of urine, without the presence of sugar in the urine. Another type of pituitary dwarfing is due to failure of other hormones, as well as the growth principle. The posterior lobe or hypothalamus may be involved, as well as the anterior lobe. The affected children are obese and frequently have voracious appetites and thirst; they are sexually under-developed, mentally backward, and very lethargic. A similar disorder resulting in great obesity without dwarfing is sometimes seen in adults.

Generalized pituitary failure produces pituitary cachexia or Simmond's disease, exhibiting marked premature senility and atrophy of the thyroid, adrenals and reproductive organs.

The Adrenal Glands

The adrenal glands are two small triangular glands lying one over each kidney, in contact with the diaphragm. They have a very rich blood supply from the aorta, the phrenic and renal arteries, and one large vein emptying into the renal vein on the left and the inferior vena cava on the right. They are supplied profusely with nerves from the sympathetic nervous system. Like the kidneys, they consist of two parts, the *cortex* and the *medulla*, which differ in structure, origin, and action.

The *cortex* gives a yellowish colour to the gland and is the outer part of it; it is closely related to the sex organs and is essential to life. If it is removed from animals they die in 10 to 15 days. The cortex is composed of lipoid material from which several crystalline steroid or fatty substances have been isolated: these are known as *corticoids* and are chemically similar to cholesterol, the bile salts and the sex hormones or corticosteroids. They are thought to be formed from cholesterol. These corticoids are divided into three groups according to their functions:

(1) Glucocorticoids,
(2) Mineralocorticoids,
(3) Sex corticoids.

The *glucocorticoids* affect the metabolism of proteins, carbohydrates and fats; particularly they increase the turning of proteins into glucose, the storing of glucose as glycogen in the liver, and the releasing of fat stored as fatty tissue for use by the tissues as fuel. They include cortisone and hydrocortisone. They affect the growth of connective tissue.

Table showing Effects of Pituitary Gland and Hormones

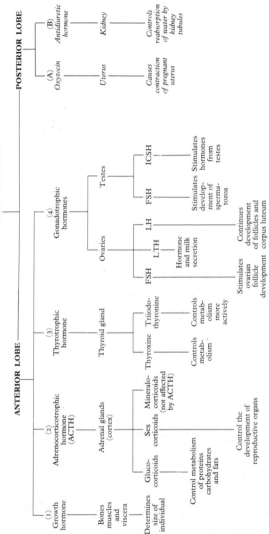

The *mineralocorticoids* include *aldosterone* which controls the reabsorption of salts by the renal tubules. When there is deficiency of this hormone, too much water and sodium are lost from the body in the urine and too little potassium is excreted, excessive quantities being reabsorbed. This results in polyuria, the passage of an abnormal quantity of urine, with possible toxic levels of potassium in the blood. In other words, the water and electrolyte balance is maintained by the mineralo-corticoids. DOC (desoxycorticosterone) is in this group and was the first of the adrenal steroids to be isolated. It is used in treatment, being first employed in the treatment of shock.

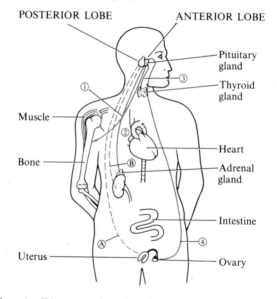

FIG. 162. Diagram to show the effects of the pituitary hormones. The figures and letters refer to the table on p. 263.

Sex corticoids include androgenic, oestrogenic and pro-gesteronic hormones which are known to affect the develop-ment and functioning of the reproductive organs and the physical and temperamental characteristics of the male and female.

Over-secretion in the young produces the infant Hercules type of child, with unusual muscular development and precocious sexual development. When occurring in women it produces masculinity, with arrest of menstruation, growth of hair on the face and alteration in voice and temperament. Another disorder produced by excessive secretion is known as *Cushing's syndrome*, in which there is an overgrowth of hair over the general surface of the body and an increase in the fatty tissue in the face and trunk, but not in the limbs. This gives the face a puffed-out, moon-like appearance.

Under-secretion, sometimes produced by tuberculous infection of the cortex, and known as Addison's disease, results in low blood pressure, marked muscular weakness, a low blood-sugar level, and bronzing of the skin and mucous membranes; there are considerable chemical changes in the composition of the blood.

The *medulla* is derived from similar tissue to the sympathetic nervous system, with which it is closely connected. It produces a very important secretion called *adrenaline*, which has the same effect on the body as stimulation of the sympathetic nervous system, except that it does not stimulate the sweat glands. In addition it secretes *noradrenaline*, which is thought to be a precursor of adrenaline, but is a distinct substance though not present in such large amounts as adrenalin. Its action is distinct. Adrenaline increases metabolism, stimulates the liver to set free stored glycogen as glucose, and causes contraction of the arterioles in the skin and internal organs, but dilates those in the muscles and heart. It also dilates the bronchial tubes, thus increasing the air intake. As a result the muscles receive an increased supply of blood, rich in glucose and oxygen.

The amount secreted varies, being increased in excitement and strong emotion such as fear or anger. As a result in these conditions extra adrenaline is present in the blood and causes a raised blood pressure with a liberal supply of fuel and oxygen in the blood stream, and an increased blood supply to the muscles and the heart. This enables the body to respond to fear by flight and to anger with fight. The adrenals are therefore sometimes called the 'glands of flight and fight' or the emergency glands. Adrenaline has other less important effects, such as contracting the little muscles which cause hair to stand on end, as can be seen in the angry or frightened dog or cat, making it appear larger to its adversary.

Noradrenaline chiefly affects the circulation, contracting the blood vessels and raising blood pressure: adrenaline given by injection raises the systolic pressure but lowers diastolic pressure; noradrenaline raises both systolic and diastolic pressure. Noradrenaline is therefore more valuable in treating low blood pressure than adrenaline.

The Thyroid Gland

The thyroid gland lies in the front of the neck. It consists of two lobes, lying one on either side, joined by a narrow band or isthmus which crosses the trachea immediately below the

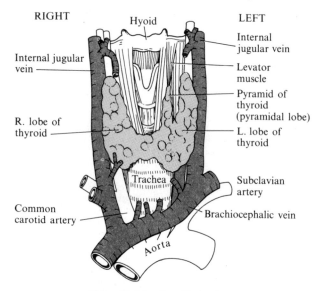

FIG. 163. The thyroid gland.

larynx. The gland is well supplied with blood vessels and consists essentially of vesicles formed by secreting cells which are grouped round little cavities called alveoli which contain a gummy substance, or colloid, containing a remarkable quantity of iodine. The cells secrete an internal secretion called thyroglobulin which is broken down to thyroxine before being released into the blood stream. Thyroxine controls the general metabolism or activity of the body tissues. Another hormone named *triiodothyronine* has now been isolated from the thyroid gland and the blood plasma. It also contains iodine and is more active than thyroxine. Thyroxine is an iodine derivative of thyronine. Thyronine itself has four iodine atoms in the molecule. The three iodine compound is even more active and the *circulating thyroxine* may be converted to this compound by the cells before it can act on the body cells. Goitre is enlargement of the thyroid gland, which used to occur in districts like Derbyshire and Switzerland, hilly or mountainous and far from the sea, where iodine is scarce. This is now prevented by adding iodine to the water or salt supply in these areas. A goitrous gland may show normal function or under- or over-secretion.

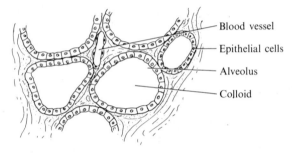

FIG. 164. Thyroid vesicles.

Under-secretion in children produces the *cretin*,⋆ a child who grows up a mentally retarded dwarf, unless treated with thyroid extract. The body is stunted, the face broad with wide nostrils, large mouth and tongue so big that the mouth cannot be closed. The abdomen is protuberant with an umbilical hernia. The child cannot be taught to walk or talk nor to have clean habits. In the adult *myxoedema* results from under-secretion; there is gradually increasing slowness in thought, speech and action, until the individual ultimately, if not treated with thyroid secretion, becomes mentally incompetent. In both conditions the skin is dry and coarse, and the hair dry, coarse, and lank. The metabolic rate in these patients is low, and they feel the cold, have a low temperature and put on weight.

⋆ The term *cretin* is derived from the French for Christian, and is used in the sense that we use innocent or simple, and perhaps also because these children were frequently cared for by members of religious orders.

Over-secretion causes thyrotoxicosis, which gives exactly the opposite picture. The individual is very excitable and nervous, the skin is fine and moist. Persons suffering from this condition feel the heat; they are very irritable and difficult to manage. They lose weight and their metabolic rate is higher than the normal. They eat well, but burn up so much fuel that they remain thin. The pulse rate is very rapid.

The Parathyroid Glands

The parathyroids are two pairs of small oval bodies the size of split peas. They lie embedded in the posterior surface of the right and left lobes of the thyroid gland. They are different from the thyroid in structure and function. Their secretion controls the level of calcium in the blood. This is important in connection with the irritability of muscle and nerve tissue.

Lack of the secretion causes a condition known as *tetany*, in which the muscles go into spasm. This may be seen after the operation for removal of the thyroid if the parathyroids have also been removed. It can be treated by giving calcium gluconate.

Over-secretion causes calcium to be lost from the bones into the blood, from whence it is excreted in the urine. The bones become softened and the blood calcium raised; there is marked depression of the nervous system, and a tendency to develop renal calculi from the excess of calcium in the blood.

The Thymus Gland

The thymus gland is an organ which lies in the lower part of the neck and chest between the lungs over the heart. It varies in size with age and grows until the child is 2 years of age; it then shrinks so that in the adult only fibrous remnants are found. When fully developed it is greyish-pink in colour and consists of two or three lobes. Its structure resembles that of a lymph node and it is, like these, associated with antibody formation. No hormone has yet been isolated from it and it may well belong to the circulation rather than the endocrine system.

The Pineal Gland

The pineal gland is a small reddish body about the size of a cherry stone, lying behind the third ventricle of the brain. Its function is unknown. In later life it becomes calcified and acts as a useful landmark in X-rays of the skull

22 The Urinary System

THE URINARY system consists of:

(1) The right and left kidneys.
(2) The right and left ureters.
(3) The bladder.
(4) The urethra.

The Kidneys

The kidneys are two bean-shaped organs lying high up at the back of the abdomen, one on either side of the spinal column, with their concave surfaces facing towards it. They lie behind the peritoneum and are embedded in a pad of fat, the perirenal fat. This pad of fat protects the kidneys from cold and injury, and together with the stout blood vessels which join them at the concave surface or hilum of the kidney, helps to maintain them in the correct position. The renal arteries enter and the renal veins and ureters leave at this point, together with nerves and lymphatic vessels.

The kidneys are a dark reddish-brown colour and measure 10 cm in length, 6 cm in width, and 4 cm in thickness in the adult. Each kidney weighs about 130 grammes. The adrenal glands lie one over each kidney. The liver on the right side and the intestines on both sides are in relation with them. The right kidney lies a little lower than the left, to make room for the liver on the right side.

The kidneys are enclosed in a *capsule* of fibrous tissue, which can easily be stripped off.

On section the kidney is found to consist of a *solid portion* and a *cavity*. The solid portion actually consists of countless minute tubules held together by connective tissue. The cavity is known as the *pelvis* of the kidney and lies at the hilum. It leads to the ureter and some anatomists call it the pelvis of the ureter.

The *solid portion* of the kidney is divided into two parts, which can be distinguished with the naked eye by the difference in their colour. These are:

(1) The *cortex* or outer part, which is reddish brown in colour.

(2) The *medula* or middle part, which is purplish in colour.

The kidney substance consists of countless minute twisted tubules called nephrons. There are over a million in each kidney. Each of the nephrons begins in a cup-shaped expansion

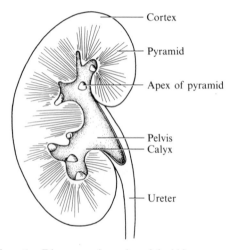

FIG. 165. Diagrammatic section of the kidney.

called the glomerular capsule, from which the tubule leads. Into the cup of each capsule comes a fine branch of the renal artery, forming a tuft of capillaries in close contact with its inner wall; the capillary tuft is called the *glomerulus*. The arteriole bringing blood to the tuft is called the *afferent vessel*, and the arteriole which carries the blood away is known as the *efferent vessel*; it is slightly smaller than the afferent vessel. The blood in the tuft is under high pressure because of this and because of its nearness to the abdominal aorta.

The convoluted tubule makes a number of twists, the proximal convolutions, on leaving the capsule and then forms a long loop, the loop of Henle, which dips down into the medulla and passes back to the cortex. The tubule next makes a distal or second series of convolutions, and ultimately empties into a straight collecting tubule of the medulla.

The efferent blood vessel which comes from the capillary tuft or glomerulus in the capsule divides to form a second set of capillaries around the walls of the convoluted tubules in the cortex. Thus the blood passes through two sets of capillaries within one organ, which does not happen in any other part of the body. The blood is collected from the second set of capillaries by small veins, which unite with other small veins to empty blood into the renal vein.

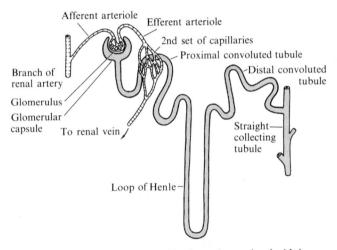

FIG. 166. A nephron and the blood vessels associated with it.

The Production of Urine

The *function* of the kidney is to secrete and excrete urine. The production of urine is a complicated process and is now more accurately understood than previously because of modern microscopic techniques and photography. It is produced by three processes, filtration, reabsorption and secretion. *Filtration under pressure* takes place from the capillary tuft, the glomerulus, where only the thin walls of the blood capillaries and of the capsule separate the blood from the kidney tubule. Huge volumes of fluid pass down the tubules from the capillary tuft continuously. The fluid is similar to the plasma in that it contains glucose, amino acids, fatty acids, salts, urea and uric acid in the same proportions, but substances with larger molecules, such as the plasma proteins and blood cells, escape only if the kidney is diseased. About 150 to 180 litres filter out in this way each day, but only about 1·5 litres leaves the body as urine.

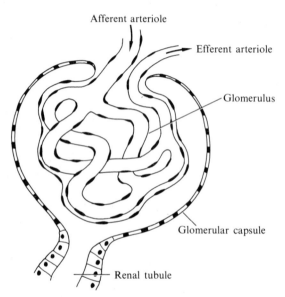

Afferent arteriole

Efferent arteriole

Glomerulus

Glomerular capsule

Renal tubule

FIG. 167. Diagram of the glomerulus and its capsule.

Selective reabsorption occurs because the cells forming the walls of the convoluted tubules are able to re-absorb the water, glucose and the salts and their ions that the body needs. All the glucose is reabsorbed in normal health. The proximal convoluted tubules absorb most of the water, salts and the glucose. As a result, instead of the large volume of fluid containing only 0·04 per cent urea, the fluid left behind in the tubules is reduced to 1·5 litres containing 2 per cent urea; it becomes acid in reaction to maintain the normal slightly alkaline reaction of the blood plasma and tissues.

Secretion occurs actively as certain substances, e.g. potassium are passed from the second capillary network into the tubule by the cells lining the tubule.

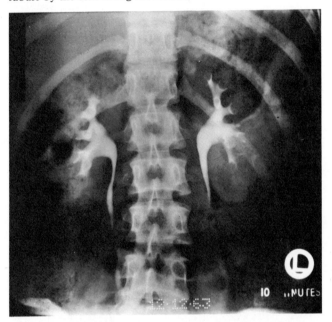

Fig. 168. A pyelogram showing the pelvis of the kidneys filled with an opaque drug which has been excreted following an intravenous injection: note the right kidney is lower than the left. Note also calyces outlining the apices of the pyramids of the medulla.

Reabsorption of water in the distal convoluted tubule is variable and is controlled by the secretion of anti-diuretic hormone from the posterior lobe of the pituitary gland. A decrease in secretion of ADH causes less water to be reabsorbed in the distal tubule, therefore more water is excreted as urine. The reabsorption of salts is controlled by the hormones of the adrenal cortex, especially aldosterone. The production of these hormones is increased or decreased according to the needs of the body to use water or salts and the electrolytes to which they give rise. Nervous control, together with the hormones adrenaline and noradrenaline, maintains the blood pressure at the high level required for filtration in the capillary tufts.

The *medulla* or inner portion of the kidney consists of straight collecting tubules into which the convoluted tubules of the cortex empty. It forms a number of cone-shaped masses which project into the pelvis of the kidney. These are called the *pyramids* of the medulla and are from eight to twelve in number.

The apices of the pyramids project into the pelvis and are covered with the mouths of the fine collecting tubules which pour the urine into the kidney pelvis. The function of the medulla is therefore to collect the urine secreted in the cortex and convey it to the pelvis.

The *pelvis* is an irregular branched cavity, which lies at the root or hilum of the kidney and leads like a funnel to the ureter. Its branches, known as the calyces of the pelvis, penetrate into the kidney substance, and each branch receives the apex of one of the pyramids of the medulla. The pyramids pour the urine into the pelvis, which conveys it to the ureters.

The Composition of Urine

Normal urine, therefore, is formed partly by filtration under pressure from the capsules and partly by reabsorption and by secretion in the tubules. It is an amber-coloured fluid varying in colour according to its quantity. It is acid in reaction and has a specific gravity of 1015 to 1025. (The specific gravity is the weight compared to the weight of an equal quantity of water, water being 1000.)

Urine consists of water, salts and protein waste products—namely, urea, uric acid and creatinine. The average composition is: water, 96 per cent; urea, 2 per cent; uric acid and salts, 2 per cent.

The percentage of urea in blood plasma is 0·04 compared to 2 per cent in the urine, hence the concentration has been increased fifty times by the work of the kidney. The salts consist chiefly of sodium chloride and also of phosphates and sulphates, produced partly from the use of the phosphorus and sulphur present in protein foods. These salts must be reabsorbed or got rid of in the quantities necessary to keep the blood at its normal reaction and maintain the water and electrolyte balance. Since this reaction and salt concentration are both essential to the

life of the blood corpuscles and the tissue cells this function of
the kidney is very important. The normal quantity of urine
secreted is 1·5 litres in 24 hr, but it is increased by drinking
and by cold weather, and decreased by reducing fluid intake
and by hot weather, exercise, and by fever since these increase
sweating. Potassium salts normally filter out and are reabsorbed or excreted as required to keep the correct level in
the body fluids. In renal failure, their excretion may be checked
so that the amount in the body fluids and tissues rises.

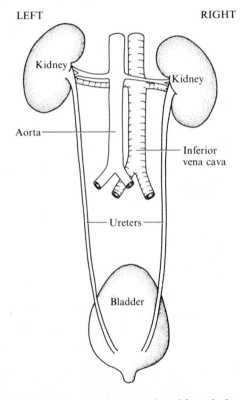

FIG. 169. The urinary system, viewed from the back.

The Ureters

The ureters are two fine tubes which run from the kidney pelvis to the bladder (see below). They are from 25 to 30 cm long and pass down the back of the abdomen and pelvis, then forward to enter the bladder low down and from the back. Their walls consist of involuntary muscle, connective tissue, and a lining of mucous membrane. The muscular coat has a peristaltic action, propelling the urine towards the bladder. If there is solid material in the urine, such as blood clot or stone, the muscle goes into spasm and causes the violent pain of renal colic.

The Bladder

The bladder is a muscular bag which acts as a reservoir for the urine, so that it is not passed continuously. It lies in the front of the pelvic cavity, immediately behind the symphysis pubis, but when full the upper part rises into the abdomen. Hence the importance of emptying the bladder before an abdominal operation, for the surgeon's knife may puncture it in operating on organs in its vicinity. Its shape and size vary according to its contents. Normally it holds 300 ml of urine when full, but it can be distended to hold very much more.

Both ureters enter and the urethra leaves the bladder at its base, the three openings making a little triangle and being about 3 cm from one another. This area, called the trigone, is very sensitive. The ureters enter obliquely, so that the muscle of the bladder wall nips them in a valve-like manner when the bladder contracts and prevents backward flow of urine through them. The exit to the urethra is guarded by a strong sphincter muscle, so that urine can only escape when it is relaxed.

The *wall of the bladder* consists of:

(1) *Peritoneum* on the upper surface only where the coils of intestine move over its surface.
(2) *Involuntary muscle*—longitudinal and circular fibres.
(3) Sub-mucous coat of areolar tissue.
(4) A lining of *mucous membrane*—transitional epithelium.

The muscle is normally in a state of mild contraction or tone.

Micturition is the passing of urine. The bladder stores the urine, which dribbles continuously into it from the ureters, until the bladder is full; the bladder then contracts and the stored urine is passed. In infants and the lower animals the act is reflex, i.e. the sensation of the full bladder is carried to the spine and there stimulates motor cells, causing the sphincter to relax. This is followed by a contraction of the bladder wall. In man the relaxation of the sphincter is under the control of the will. The sense of fullness of the bladder causes the desire to pass urine, but relaxation of the sphincter can be controlled till a convenient opportunity occurs. It is not, however, quite similar to the passing of faeces, for the bladder continues to become more and more distended, and the sensation of fullness more and more acute, till control breaks down and urine is passed, relieving the tension and back pressure on the kidneys.

In disease micturition may be affected in many ways. The sphincter may be paralysed in a state of contraction (spastic paralysis), so that it cannot be relaxed. This will cause *retention* of the urine and the bladder will become overfull. If this is not relieved by catheterization, distension will continue and will distend the orifice which the sphincter guards, allowing a little urine to dribble away continuously though the bladder still remains full. This is called *retention with overflow*. This is bad for both bladder and kidney, causing loss of tone and interference with blood supply to the stretched bladder wall and back pressure on the kidney. It should never be allowed to occur.

The sphincter may be paralysed in a state of relaxation. In such conditions urine will dribble constantly from an empty bladder, as the bladder cannot retain it. This is comparatively rare.

In other cases of nervous disease and in anaesthesia the control of the brain over micturition may be abolished, so that the act becomes again reflex as in the lower animals; the bladder fills and empties reflexly without the individual being aware of any sensation of fullness or able to control the relaxation of the sphincter.

The Urethra

The urethra is a canal leading from the bladder to the outside skin. It is 2·5 to 4 cm long in the female and runs straight downwards and forward from the base of the bladder to the external skin in the vulva. It is lined with mucous membrane. In the male it is 15 to 20 cm long and runs in a curved course through the prostate gland, loose connective tissue, and the penis. It opens on to the external skin at the tip of the penis normally. The male urethra is the common passage for urine and the semen or reproductive fluid (see p. 242).

23 The Nervous System

THE NERVOUS system is the system of communication between the various parts of the body. It is, in fact, not unlike the telephone system, with the brain comparable to the central switchboard and operators, the spinal cord like the main cable, and the nerves like the telephone wires, ending in receivers and dischargers of messages in the body tissues. Like the skeletal and muscular systems, the nervous system is made of a special tissue, nervous tissue, which forms this particular system. Nervous tissue consists of *nerve cells* and *nerve fibres*, each cell having one long fibre and several short fibres growing from it, and all the fibres arising from a cell. The cell and its fibres are together called a *neurone*, and the neurone is the unit of which the system is built up. The cells and fibres are bound together by a special type of connective tissue known as *neuroglia*, which binds them into a solid, though a very soft and delicate tissue (see Fig. 32).

The long fibre is called the *axon*, and the short fibres are called the *dendrites* of the cell. The dendrites pick up nerve impulses or messages and carry them to the cell. The axon carries impulses from the cell, and at its ending passes them on to another nerve cell or dendrite, or to some other tissue, such as muscle or gland tissue; for example, a nerve cell in the spinal cord, picking up impulses by its dendrites, will pass the impulses out to a muscle and produce contraction, resulting in movement.

The *nerve cells* are grouped together to form the *grey* matter of the nervous system. Grey matter is found at the periphery of the brain, the centre of the spinal cord, and in ganglia. A ganglion is a small isolated mass of nerve cells. The *nerve fibres* are grouped together to form the *white* matter of the nervous system. White matter is found at the centre of the brain, the periphery of the spinal cord and in the nerves. A nerve is simply a bundle of nerve fibres bound together by connective tissue. The whiteness of a nerve fibre is due to its sheath, which makes a marked difference in the colour of the tissue, readily obvious to the eye.

The nerve sheath consists of two coats:

(1) An outer coat of connective tissue called the *neurilemma*.

(2) A thick inner fatty sheath, the *myelin sheath*. This fatty sheath is interrupted at intervals, and here the outer neurilemma dips in and forms notches called the *nodes of Ranvier*.

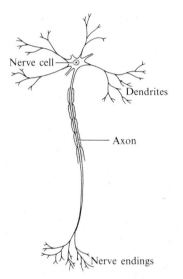

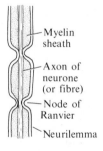

Nerve cell

Dendrites

Axon

Myelin sheath

Axon of neurone (or fibre)

Node of Ranvier

Neurilemma

Nerve endings

FIG. 170. A typical neurone, with details of the nerve sheath enlarged.

The *functions of the myelin sheath* are thought to be:

(1) *Protection* from pressure and injury.

(2) *Nutrition*. As the fibres may be of great length, varying from 2·5 to 75 cm or more the distal part of the fibre may be a long way from the cell which controls the nutrition of its protoplasm. Nerve cells in the lumbar region of the spinal cord give off nerve fibres to the muscles of the foot and may be more than a metre in length. Hence it is suggested that the sheath may help to nourish the fibre.

(3) *Insulation.* It is suggested that the sheath acts like the casing of an electric wire, so that the impulses carried by the nerve are not transmitted to adjacent nerves or tissues, except through the end of the fibre.

The nerve cell and its fibre are one living unit, and if the fibre is cut off from the cell it will die, as it is the cell which contains the nucleus, and the protoplasm of the fibre cannot live if cut off from this. On the other hand, the cell and the portion of the fibre still attached to it remain alive, and a new fibre can grow from the cut end provided the neurilemma remains intact. The fibre will not grow through the fibrous scar tissue of a wound, but it will grow along the old nerve sheath. As a result the nerve supply can be restored if a nerve is cut, though it will take time for the nerve fibre to grow, as these fibres, though often short, can sometimes be as much as a yard in length. It will, however, take much longer for the newly-grown fibre to learn to function fully. On the other hand, if the nerve cell is destroyed by injury or disease, the fibre will also die and neither can be replaced.

Neurones are of two main types:

(1) *Efferent neurones,* which carry impulses *out* from the brain to the tissues, e.g. (a) *motor neurones,* which supply the muscles and produce movement; and (b) *secretory neurones,* which supply the glands and produce secretion.

(2) *Afferent neurones,* which carry impulses *in* to the brain from the tissues. These give rise to sensations such as touch, pain, heat or cold, and are also known as *sensory neurones.*

If any part loses its nerve supply, the condition of the tissues becomes unhealthy; sores are likely to occur, do not heal readily, and are very liable to septic infection, until the nerve supply to the part is re-established.

There are also *association* or *connector neurones* which run between the afferent and efferent neurones in the brain and spinal cord, linking them together. These 'connector' neurones are like the internal telephones in a large building such as a hospital: telephone wires come in and out of the building like the afferent and efferent neurones; there are also internal telephone wires between the different wards and departments, so that all can work smoothly together.

The neurone described above is the typical neurone, consisting of a cell giving off many dendrites and one axon. There are other varieties, and these are found in the sensory neurones

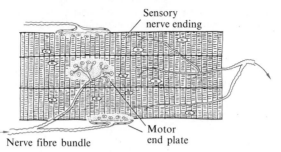

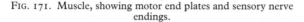

FIG. 171. Muscle, showing motor end plates and sensory nerve
endings.

running from the tissues of the body into the cord or brain. If
the typical neurone was found here the cell would lie in the
tissue—for example, in the skin—where the stimuli would be
picked up by the dendrites and the fibre would run into the
spinal cord or brain and there pass on its impulse to another

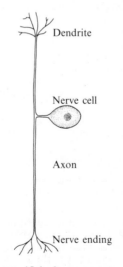

FIG. 172. Unipolar neurone.

nerve cell. This would be unsatisfactory, as nerve cells are extremely soft and delicate and the nerve cells would be destroyed by pressure or injury to which the skin is exposed. Since the fibre is merely an out-growth from the cell it would also die, as the cell nucleus is necessary for the life of the proto-plasm of the whole neurone, and sensation would be lost. These neurones, therefore, have a different structure; they have specialized nerve 'endings' in the skin by which sensa-tions are picked up. The fibre runs from these endings into the vertebral column, and there, protected by bone, the delicate cell is found joined to the fibre by a T-like connection; from this cell the fibre runs on into the cord to form the normal endings, which pass on the impulse to the afferent and associa-tion nerve cells of the cord (see Fig. 182). These neurones are called unipolar neurones.

The axon ends as a rule in a tree-like branching by which the stimulus carried by the axon is passed on to the dendrites of another nerve cell. Where the fibre ends in nervous tissue and passes the impulse on to another nerve cell the axon must communicate with the dendrites of another nerve cell for the passage of the impulse to take place; this point is called the *synapse*. It permits passage of an impulse in one direction only and offers some resistance to its passage, but the dendrites of the adjacent neurones do not actually touch the axon. The impulse passes across the synapse because a chemical sub-stance, which in many cases is acetylcholine, is released at the junction, which then stimulates the dendrites of another neurone or neurones. Here delay may occur.

The fibre endings in the tissues of the body form the *nerve endings*. These nerve endings are of two types:

(1) *Motor* nerve endings, which give off stimuli to the muscles, producing contraction; these are called motor end plates. Each muscle fibre is stimulated through a single motor end plate.

(2) *Sensory* nerve endings, which pick up impulses and carry them into the spinal cord and the brain; these are the endings of afferent or sensory neurones, which give rise to sensations, but many remain below the level of consciousness.

These *nerve endings* in the tissues are often peculiar in struc-ture. Some of them are simple tree-like branchings similar to the fibre endings found in the nerve tissue. On the other hand, motor nerve endings branch, but the tip of each branch carries a minute disc shaped plate in the actual muscle fibre; these are the *motor end plates*. Similarly the nerves of special sense have each their characteristic endings. The nerves of touch in the skin end in little round bodies which, on section, look rather like an onion under the microscope. These are known as the touch or tactile corpuscles, which are stimulated by pressure. The nerves of sight end in cone- or rod-shaped cells in the retina of the eye which are stimulated by light. The nerves of hearing end in hair-bearing cells in the ear, which are affected by sound. The nerves of taste end in taste buds in the tongue.

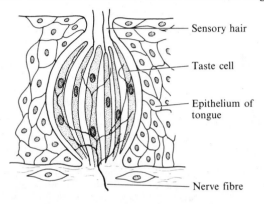

Sensory hair

Taste cell

Epithelium of tongue

Nerve fibre

FIG. 173. A taste bud in the tongue.

The nature of the impulses carried by the nerve tissue is electric: but as in the electric battery chemical reactions are associated with the passing of the electric current, so in the body chemical changes associated with the sodium, potassium and chloride-ions accompany the electric impulse. The functioning of nerve tissue has therefore both an electrical and chemical factor.

The *nervous system* is divided into:

(1) The **central nervous system**, which controls the voluntary muscles of the head, trunk and limbs, and is responsible for all movement in them and for all sensation in skin, muscles, bones and joints and the special sense organs. This supplies nerves to the limbs and to the structures which form the walls of the cavities of head and trunk, and the covering muscles and skin.

(2) The **autonomic nervous system**, which controls all involuntary muscle. This supplies all the internal organs,

and is for the most part made up of efferent neurones supplying the glands and the muscular walls of the internal organs and the blood vessels, as the internal organs have comparatively little sensation. There are receptor nerve endings which are stimulated by chemical changes, or by pressure on and stretching of the tissues which form these organs.

Central Nervous System

The central nervous system consists of brain, spinal cord, and the nerves given off by these to the outer parts of the body, and called the peripheral nerves to distinguish them from the nerves supplying the internal organs (autonomic nervous system).

The central nervous system develops very early. When the developing fertilized ovum consists merely of a mass of cells arranged in an outer, middle and inner layer, a groove appears on the surface. This groove is called the neural groove. The outer layer of cells in which it is formed also develops into the skin, while the middle layer forms the muscle, bone and other connective tissues of the body, and the inner layer forms only the lining membranes of hollow internal organs, e.g. the alimentary canal and the air passages. The neural groove is gradually turned into a canal, its walls growing up and joining round it. The walls of the canal develop to form the spinal cord, which has always a tiny canal running through the centre of it. At the upper end, where the head is to develop, the canal becomes enlarged and the walls of the enlarged cavity develop to form the brain. Although the groove forms on the surface of the embryo, the other layers gradually grow up around it, so that brain and cord become internal organs, well protected from injury, before birth.

The Brain

The brain, when fully developed, is a large organ filling the cranial cavity. It weighs about 1·4 kg. Its large size, compared to the total weight of the body, is one of the main anatomical differences between man and the other animals. The brain is a complicated organ. Early in its development, its cavity becomes divided by constrictions into three parts, the forebrain, the midbrain and the hindbrain. From the walls of these three parts, the fully developed brain is formed, the cavities persisting, and being called ventricles. The brain consists of five main parts:

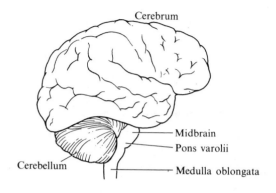

FIG. 174. The brain, showing parts.

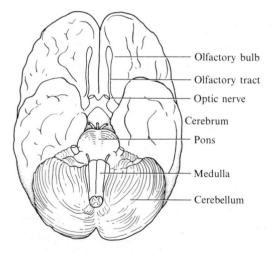

FIG. 175. The brain, viewed from below.

1. The *cerebrum*
2. The *midbrain*
3. The *pons varolii* ⎱ The brain stem
4. The *medulla oblongata* ⎰
5. The *cerebellum*

The midbrain, pons varolii and medulla oblongata are together often called the *brain stem*, as they are comparatively small, and occupy the back lower part of the cranial cavity only, the cerebrum completely overlapping them. In the lower animals the cerebrum is comparatively small and undeveloped, the other parts forming the greater part of the brain. In the higher types with more intelligence the cerebrum becomes progressively larger, reaching its greatest size in man. The cerebrum of the pig, for example, is about the size of man's little finger.

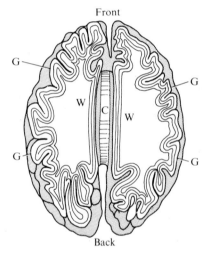

Front

Back

FIG. 176. Section through the cerebrum, viewed from above. G, the grey matter of the convoluted surface; W, the white matter forming the central portion; and C, the corpus callosum, the bridge of white matter joining the two hemispheres of the cerebrum.

The *cerebrum* fills all the vault of the cranium from the level of the eyebrows in front to the occiput at the back. It is divided into two *hemispheres*, right and left, by a deep central fissure (longitudinal cerebral fissure). These hemispheres control the opposite sides of the body, so that disease of the right side of the cerebrum affects the left side of the body, and vice versa. Each hemisphere contains a small cavity, or ventricle, known as the right and left lateral ventricles.

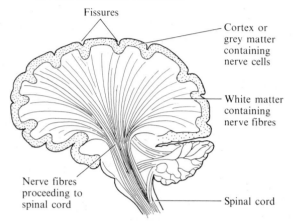

Fissures

Cortex or grey matter containing nerve cells

White matter containing nerve fibres

Nerve fibres proceeding to spinal cord

Spinal cord

FIG. 177. Diagrammatic section of the brain, to show grey matter on the surface and white matter in the centre. The association fibres in the cerebrum are not shown.

Each hemisphere is subdivided into *lobes*, which correspond roughly with the bones of the cranium, the chief lobes being:

(1) The frontal lobe.
(2) The parietal lobe.
(3) The temporal lobe.
(4) The occipital lobe.

These are divided from one another by deep fissures, known as sulci, e.g. the central sulcus divides the frontal and parietal lobe, the lateral sulcus divides the temporal lobe from the frontal and parietal lobes and the parieto-occipital sulcus passes downwards and forwards from the back of the hemisphere (see Fig. 178).

The cerebrum consists of grey matter or nerve cells on the surface and white matter or nerve fibres in the centre. The surface is convoluted, i.e. thrown into ridges and depressions, which greatly increase its extent and therefore increase the amount of grey matter or nerve cells, which are the seat of our mental capabilities. The convolutions of man's brain are more extensive than those of the lower animals, which further increases the possibility of mental development. The grey matter on the surface or cortex of the cerebrum contains many important nerve centres. A nerve centre is a group of nerve cells with similar function, e.g. the centre of sight is a group of sensory nerve cells linked by nerve fibres to the eye through which we see; a motor centre is a group of motor cells linked by nerve fibres to the muscles of a certain area; through these nerve cells and fibres the muscles are stimulated to contract and produce movement.

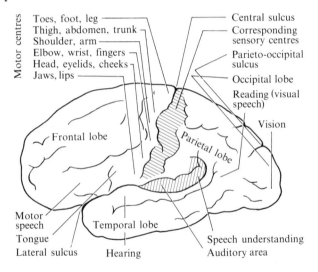

FIG. 178. The cerebrum, showing the lobes and the main nerve centres.

The functions of the cerebral cortex. The cerebrum contains many important nerve centres which make it not only the largest but the most highly developed part of the brain. It contains:

(1) The *motor* centres, controlling all the voluntary muscles of the muscular system.

(2) The *sensory* centres, which give sensation to the skin, and to a lesser extent to the muscles, bones and joints.

(3) The centres of *special sense*: sight, hearing, smell, taste and touch.

(4) The centres of the *higher mental powers*, e.g. consciousness, memory, intelligence, reasoning power, etc.

The position of the centres in the lobes of the cerebrum is to a large extent known. The motor centres lie in the frontal lobe in front of the central sulcus. Those for the lower limbs are at the top; next come those of the trunk, upper limbs, neck and head, extending outwards and downwards in front of the fissure. The motor centres for the eyes lie farther forward (hence frontal headache may be due to eye trouble). The speech centre also lies in the frontal lobe: usually in the left cerebral hemisphere.

The sensory centres for the limbs, trunk and head lie in the parietal lobe just behind the central sulcus in positions corresponding to the motor area. The centre of sight is in the occipital lobe. The centres of smell and hearing are in the temporal lobe.

The *white matter* consists of nerve fibres running to, from, and between the cells of the cortex. There are three types of fibres:

(1) *Motor fibres* running from the motor centres of the cortex out through the base of the brain into the spinal cord; these carry impulses from the brain.

(2) *Sensory fibres* running in from the base of the brain to the sensory centres of the cortex; these bring impulses to the cerebral nerve centres.

(3) *Association* or *connector fibres*; these run from one nerve centre to another and from hemisphere to hemisphere, linking them to one another, so that the various centres of the brain can work as a whole and communicate one with another. A mass of these connector neurones, made up of fibres linking the two hemispheres to one another, forms a bridge-like structure at the base of the fissure which separates the hemispheres from one another. This is called the *corpus callosum*.

The brain is comparable to a very complicated telephone exchange, which has wires bringing messages in to certain rooms and wires taking them out from other rooms, and also an internal telephone, so that each room in the exchange is linked to others, and messages can pass between them. The nerve centres are like the rooms receiving and discharging messages, and the nerve fibres are like the wires bringing in messages to the nerve centres, carrying them from centre to centre, and sending messages out to the spinal cord and tissues. The nerve centres not only receive messages: they store them up so that they can be recalled, giving rise to memory, and can be used as a guide to the action to be taken when a similar stimulus is received again, resulting in intelligent reaction to sensory stimulus. For example, a child seeing a silver teapot in use may touch it, attracted by its bright surface. He will feel pain as a result of the sensory stimuli from the hand. He will remember that the teapot has burnt his hand and avoid touching it again.

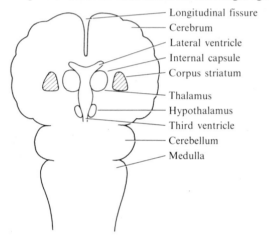

FIG. 179. Cerebral nuclei.

The nerve fibres run towards a point at the centre of the base of the cerebrum, where they form two stalks which pass into the midbrain. Between them in the base of the cerebrum is a small central cavity, the third ventricle into which the lateral ventricles lead. There is also here a small amount of grey matter, which consists almost entirely of sensory cells. These cells receive the stimuli which are being brought into the brain by the afferent nerve fibres, and relay them on by fresh neurones to the nerve centres of the cortex. This grey matter is sometimes spoken of as the cerebral, or basal, nuclei, but its various parts have each their own names.

The cerebral nuclei include the thalami on either side, forming the lateral walls of the third ventricle. These are made up mainly of sensory cells relaying sensory stimuli to the cortex.

The corpus striatum lies lateral to the thalamus on either side: it is separated from it by the internal capsule which consists of the converging tracts of motor and sensory nerve fibres between the thalamus and the corpus striatum and running to and from the cerebral cortex. The corpus striatum is so called because nerve fibres pass through it, making it appear striped. This is largely a motor relay station. The pyramidal tract and many other motor and sensory tracts run in the narrow internal capsule. This is the area usually involved when a patient has a stroke. The hypothalamus exercises an influence over the autonomic nervous system; it contains the heat regulating centre and communicates with the posterior lobe of the pituitary gland.

The *midbrain* consists of two stout stalk-like bands of white matter which pass out from the base of the cerebrum and run into the pons varolii. It contains some grey matter, and a fine canal, called the aqueduct of the midbrain, runs through it leading from the third ventricle above to the fourth ventricle, which lies between the cerebellum behind and the pons and medulla in front. The white matter consists of motor and sensory fibres running from and to the nerve centres of the cerebral cortex and the cerebral nuclei.

The *cerebellum* is smaller than the cerebrum and lies below it at the back. It is sometimes spoken of as the hindbrain. It is similar in structure to the cerebrum, being divided into two hemispheres, consisting of grey matter on the outside and white matter in the centre, and having a finely convoluted surface. There are three pairs of cerebellar stalks or peduncles joining it to the midbrain above, the pons in front, and the medulla below. Impulses passing to the cerebellum via these peduncles keep it informed concerning the state of the muscles. Its *functions* are not completely understood, but it is known to be responsible for maintaining *equilibrium*, muscle *co-ordination* and muscle *tone*. Destruction of the cerebellum by disease results in loss of the power to co-ordinate muscular actions, and therefore the patient's movements are exaggerated and awkward, e.g. a full cup cannot be lifted and drunk without spilling the liquid. The patient cannot stand or walk steadily, but staggers and moves like a drunken man. Everything happening in the cerebellum is below the level of consciousness.

The *pons varolii* consists almost entirely of white matter, and forms the link joining the various parts of the brain to one another. It consists of two main parts:

(1) A bridge-like portion joining one hemisphere of the cere-

bellum to the other, formed of fibres running from one hemisphere to the other; this gives the part its name, as *pons* is the Latin for bridge. (This is the middle cerebellar peduncle.)

(2) A mass of fibres running from the cerebrum and midbrain above under the bridge-like portion to the medulla.

Two stout stalk-like bands of white matter pass out from the base of the cerebrum, consisting of the motor and sensory fibres running from and to the nerve centres of the cortex. These bands of white matter are a continuation of the nerve fibres which pass through the internal capsule and pass through the midbrain and pons, filling in the gap under the bridge-like portion completely, and run on into the medulla or cerebellum below.

The pons therefore joins the cerebrum above through the midbrain to the medulla and cerebellum below, and also joins the two hemispheres of the cerebellum to one another.

The *medulla oblongata* joins the pons above to the spinal cord below, being the link between brain and cord. It is often called the spinal bulb, as it is similar to the cord in structure, but slightly thicker. It lies just within the foramen magnum at the base of the brain.

It consists of white matter on the surface and grey matter in the centre, as does the cord. The white matter consists of:

(1) Efferent or motor fibres running out from the brain to the cord.

(2) Afferent or sensory fibres running in from the cord to the brain.

Most of the motor fibres cross here as the decussation of the pyramids so that the left side of the brain controls the right side of the body, and vice versa. The grey matter contains:

(1) The *vital centres*, i.e. the respiratory, cardiac and vasomotor centres, which are essential to the continuance of life. Conscious life is dependent on the higher centres of the cerebrum. These may be thrown temporarily out of action, as during sleep and unconsciousness, but the lower centres of the medulla must continue to function, or death at once ensues. Injury here therefore causes instant death.

(2) *Reflex centres* controlling the food and air passages; such reflex actions as swallowing, vomiting, coughing and sneezing are produced by these centres in the medulla.

The brain is not solid, but contains four cavities known as the ventricles of the brain. There are two lateral ventricles, one in either hemisphere of the cerebrum, while the third lies in the midline at the base of the cerebrum, and the fourth is between the pons, the medulla oblongata and the cerebellum. These ventricles communicate with one another and with the subarachnoid space, and contain cerebrospinal fluid.

The meninges. The brain and cord are covered by membranes known as the *meninges*. There are three of these:

    (1) The dura mater.
    (2) The arachnoid mater.
    (3) The pia mater.

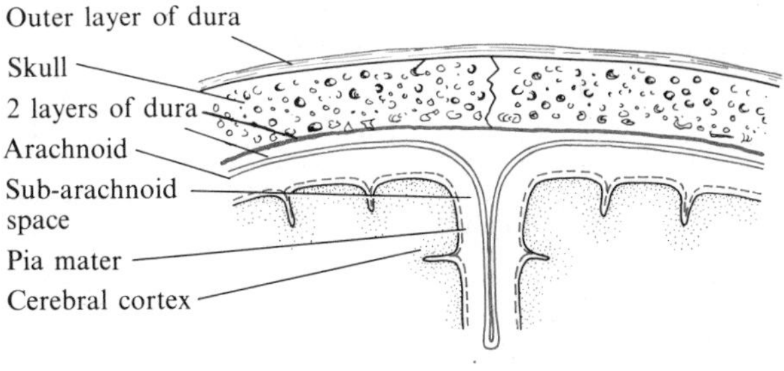

FIG. 180. Diagram of the meninges.

The *dura mater* is a tough fibrous membrane composed of two layers. It is attached to the inner surface of the skull, forming the periosteum, but the two layers are separate where they enclose a venous sinus. At the foramen magnum the outer layer continues as the periosteum on the outer surface of the skull, and only the inner layer passes down the spinal cord. One portion dips in between the cerebrum and cerebellum (the *tentorium cerebelli*), another portion dips down between the two hemispheres of the cerebrum (the *falx cerebri*). These help to support and protect the delicate brain substance.

The *arachnoid mater* is a delicate serous membrane lying close beneath the dura and dipping down with it between the main portions of the brain. The name arachnoid means like a spider's web, and is given because of the microscopic appearance of the tissue.

The *pia mater* is a delicate membrane, richly supplied with blood vessels, which covers the actual surface of the brain and dips down into all its convolutions. It carries the blood supply to the underlying brain.

The *cerebrospinal fluid* is a clear watery *fluid* which collects between the arachnoid and pia mater in the subarachnoid space, into which it oozes from the choroid plexus in the ventricles of the brain. It protects the brain and spinal cord, acting as a water cushion between delicate nerve tissue and the bony walls of the cavities in which these structures lie. It also nourishes and cleanses, washing away waste and toxic substances. It is absorbed by the arachnoid granulations in the venous sinuses.

The brain is thus protected by the hair, the tough scalp, the cranial bones, the dura mater and the cerebrospinal fluid.

The Spinal Cord

The spinal cord is a cylinder of nerve tissue about the thickness of the little finger and 38 to 45 cm long. It joins the medulla oblongata above and runs down the spinal canal to the level of the base of the body of the first lumbar vertebra, where it ends in a bunch of nerves called the *cauda equina* (which is the

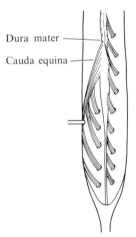

Dura mater

Cauda equina

FIG. 181. The cauda equina.

Latin for horse's tail). These nerves pass out from the lumbar and sacral regions of the vertebral column. The cord gives off nerves in pairs throughout its length. It varies somewhat in thickness, swelling out in both the cervical and lumbar regions, where it gives off the large nerve supply to the limbs. These are called the cervical and lumbar enlargements. The cord is deeply cleft back and front, so that it is almost completely divided into right and left sides like the cerebrum.

The structure of the cord. The cord, like the medulla, consists of white matter on the surface and grey matter in the centre. The white matter consists of fibres running between the cord and the brain only, not to the body tissues. It contains:

(1) *Motor* or *efferent fibres* running down from the motor centres of the cerebrum and cerebellum to the motor cells of the cord.

(2) *Afferent* or *sensory fibres* running up the cord from the sensory cells of the cord to the sensory centres of the brain.

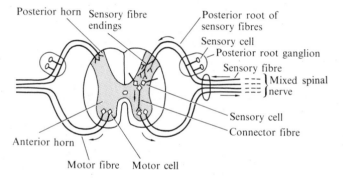

FIG. 182. A section of the spinal cord, showing central grey matter and neurones. Sensory fibres of heat and pressure cross the cord.

The *grey matter*, on cutting across the cord, has an H-shaped pattern, with two portions projecting forwards, one on either side, called the *anterior horns*, and two portions projecting backwards, one on either side, called the *posterior horns*. The anterior horns consist of motor cells which give off motor fibres to the muscles of trunk and limbs. These fibres run out from the cord and form the motor fibres of the spinal nerves. A motor impulse from the motor centre in the brain is brought by an efferent fibre down to the motor cell in the cord. The motor cells in the brain and the fibres which run from them to the anterior horn of the cord are called the upper motor neurones; the motor cells in the cord and the fibres which run from them out to the muscles are called the lower motor neurones.

The posterior horns consist of sensory cells, which give off *sensory fibres*, which run up the cord to the centres at the base of the brain; these pass on the impulses they receive to the sensory centres of the cerebrum. Sensory fibres from body tissues run into and end in the posterior horn of the cord, bringing in impulses from the tissues to the sensory cells and to the connector cells of the posterior horn. Impulses are brought in by afferent fibres from the tissues to the cells of the cord, which pick up the impulses and pass them up the cord to the sensory centres of the brain, producing sensation. Sensations of temperature and pain cross at once, to the opposite side, the others at higher levels.

In the cord, as in the brain, association or connector fibres run between the sensory and motor cells and carry impulses between them. As a result, an impulse brought into the posterior horn may be passed on to cells in the anterior horn of the cord on the same or another level, producing reflex action.

The functions of the cord. The spinal cord has two main functions:

(1) It is the link between the brain and the nerves supplying the outer parts of the trunk and the limbs.

(2) It is a great centre of reflex action (see p. 304).

The Peripheral Nerves

Nerves are given off to the tissues from both the brain and the cord. There are:

(1) Twelve pairs of *cranial* nerves, arising from the brain.

(2) Thirty-one pairs of *spinal* nerves, arising from the spinal cord.

The *cranial nerves* supply the organs of the head and neck for the most part. They are, as a general rule, either motor or sensory nerves, but some are mixed nerves. They include the nerves of smell, taste, sight (the optic nerve), and hearing (the vestibulocochlear (auditory) nerve) and the vagus nerve.

The vagus nerve supplies organs in the neck, thorax and abdomen. Its name comes from the same source as the word vagabond, meaning wanderer, as, unlike the other cranial nerves, its branches wander throughout the trunk. It helps to form the autonomic nervous system (see p. 306).

The trigeminal nerve is so called because of its three large sensory branches which come off from a ganglion lying within the cranial cavity called the trigeminal ganglion. It is this nerve which is affected in cases of facial neuralgia, the cause of this often being neglected septic teeth.

The cranial nerves are:

Name	Type	Function and Distribution
1. Olfactory nerve	Sensory	Supplies the nose and is the nerve of smell
2. Optic nerve	Sensory	Supplies the eye and is the nerve of sight
3. Oculomotor nerve	Motor	Supplies the muscles moving the eye and those within it
4. Trochlear nerve	Motor	Supplies one of the muscles moving the eye (superior oblique)
5. Trigeminal nerve	Motor and sensory	Supplies the muscles of mastication and gives off three sensory branches to the orbits, upper and lower jaws respectively (the ophthalmic, maxillary, and mandibular nerves)
6. Abducent nerve	Motor	Supplies one of the muscles moving the eye (lateral rectus)
7. Facial nerve	Motor and sensory	Supplies the muscles of facial expressions and taste to the anterior part of the tongue
8. Vestibulo-cochlear nerve	Sensory	Supplies the ear and is the nerve of hearing and balance
9. Glosso-pharyngeal nerve	Mixed	Supplies the posterior part of the tongue and pharynx and is the nerve of taste. Supplies motor fibres to the pharynx
10. Vagus nerve	Mixed	Supplies the internal organs controlling both secretion and movement. It supplies the larynx
11. Accessory nerve	Motor	Supplies the muscles of the neck. It supplies most of the muscles of the pharynx and the soft palate, and one branch joins the vagus to supply the pharynx
12. Hypoglossal nerve	Motor	Supplies the tongue muscles

The facial nerve is the nerve affected in cases of facial paralysis. It leaves the cranium with the vestibulo-cochlear nerve and runs through the ear, so it may become infected or injured in diseases of and operations on the ear.

When the vagus and accessory nerves are affected, the muscles involved in swallowing are paralysed and there is great danger of asphyxia, as neither the natural secretions nor food and drink can be swallowed and are liable to be inhaled into the air passages. This occurs particularly in the bulbar form of anterior poliomyelitis. It may cause complete blockage of the airway or 'aspiration pneumonia'.

The *spinal nerves* supply the muscles of the trunk and limbs with the power of movement, and give sensation to the skin and to a lesser extent to the muscles, bones and joints of these parts. There are thirty-one pairs of spinal nerves. They are mixed nerves, containing both motor and sensory fibres. They arise from the spinal cord by two roots:

(1) The *anterior* or *motor root*, consisting of fibres coming from the motor cells of the anterior horn and passing out to the muscles.

(2) The *posterior* or *sensory root*. This consists of the afferent or sensory fibres, bringing in sensory stimuli from the skin, and other tissues to a lesser extent. These fibres are the ones described on p. 282. Their endings, by which stimuli are picked up, are found in the skin. From these endings the fibres run into the vertebral canal and form the posterior roots of the spinal nerves. On each root is a ganglion, which is visible as a marked swelling, and is called the posterior root ganglion. This consists of the cells of these fibres, which thus have the bony protection of the spinal column, and are joined by a T-shaped branch to the fibre. From the ganglion the fibre runs on and enters the posterior horn of the spinal cord, where it ends in a widely spread tree-like branching, which passes on the stimuli brought in by the fibre to the sensory cells of the cord (see Fig. 187), on the same or the opposite side.

The spinal nerves are divided into groups according to the region of the cord from which they arise. There are:

(1) Eight pairs of *cervical* nerves, one above the atlas and one below each of the cervical vertebrae

(2) Twelve pairs of *thoracic* nerves, one below each thoracic vertebra

(3) Five pairs of *lumbar* nerves ⎫
(4) Five pairs of *sacral* nerves ⎬ Derived from the cauda
(5) One pair of *coccygeal* nerves ⎭ equina

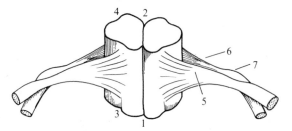

FIG. 183. A section of the spinal cord, showing a pair of spinal nerves arising from it. 1 and 2, fissures which divide the right and left sides of the cord; 3 and 4, smaller fissures at the site from which the nerves arise; 5, the anterior motor root of the spinal nerve; 6, the posterior or sensory root, on which the posterior root ganglion (7) shows as a swelling close to the point where the two roots join.

The spinal nerves give off short posterior branches, which supply the muscles of the back of the neck and trunk, and long anterior branches, which provide the nerves of the limbs and the sides and front of the trunk.

In certain regions these nerves branch immediately they leave the spinal canal, and the branches join up with one another to form the nerves supplying the various muscles and parts. This interbranching is called a plexus. Plexuses are formed in all regions except the thoracic region.

The *cervical nerves* form two plexuses:

(1) The *cervical plexus*, which supplies muscles of the neck and shoulder, and also gives off the *phrenic* nerve supplying the diaphragm.

(2) The *brachial plexus*, which supplies the upper limb.

The *brachial plexus* gives off three main nerves—the radial, median and ulnar nerves. The *radial nerve* runs round the back of the humerus and down the outer side of the forearm. It supplies the extensor muscles of the elbow, wrist and hand. It is exposed to pressure in the armpit and against the humerus, and injury to it is the cause of drop-wrist, when these joints are flexed and cannot be extended. This may result from the use of badly padded, cheap crutches, with no hand rest, pressing in the armpit or from pressure by the edge of the operation table on the nerve against the humerus, if the patient's arms are allowed to hang down during an operation.

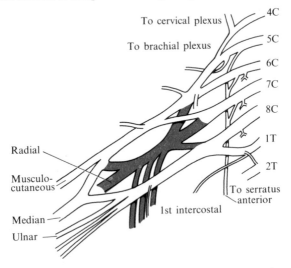

FIG. 184. The brachial plexus and the main nerves arising from it.

The *ulnar* and *median nerves* run down the inner side and middle of the limb respectively, and supply the flexor muscles of the wrist and hand. Injury to them causes hyperextension and gives rise to the 'claw-hand', the unopposed extensor muscles coming into play. A fourth smaller nerve, the musculo-cutaneous nerve, supplies the flexors of the elbow-joint, the biceps, and brachialis muscles. It is the ulnar nerve, crossing in the groove between the back surface of the internal epicondyle of the humerus and the olecranon, which is knocked when we say we have 'knocked our funny-bone', the tingling pain passing down the nerve to the hand.

The *thoracic nerves* supply the muscles of the chest and the main part of the abdominal wall.

The *lumbar nerves* form the lumbar plexus, which gives off one main nerve, the *femoral nerve*. This runs down beside the psoas muscle under the inguinal ligament into the front of the

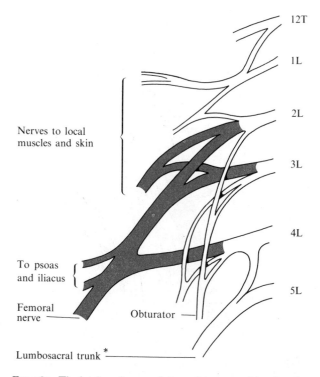

FIG. 185. The lumbar plexus and the main nerves arising from it.
* This trunk joins the sacral plexus.

thigh and supplies the muscles there. The lumbar plexus also gives off branches to the lower abdominal wall.

The *sacral nerves* with branches from the fourth and fifth lumbar nerves form the sacral plexus, which gives off one large nerve, the *sciatic nerve*. This is the largest nerve in the body. It leaves the pelvis by the sciatic notch, runs across the back of the hip-joint and down the back of the thigh, supplying the muscles there. It divides above the knee into two main branches:

(1) The *peroneal* nerve, which supplies the muscles of the front of the leg and foot.

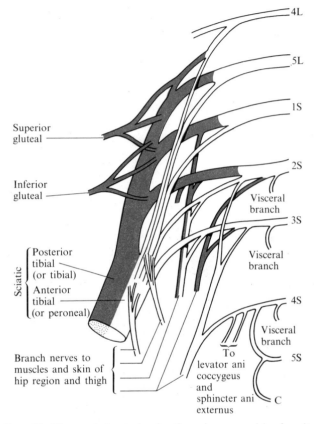

FIG. 186. The sacral plexus, showing the main nerves arising from it.

(2) The *tibial nerve*, which supplies the muscles of the back of the leg.

The sciatic nerve therefore supplies the whole of the leg below the knee except for a small sensory branch from the femoral nerve.

The *coccygeal nerves*, with branches from the lower sacral nerves, form a second small plexus on the back of the pelvic cavity, supplying the muscles and skin in that area, e.g. the muscles of the perineal body, the external sphincter of the anus, the skin, and other tissues of the external genitals and the perineum, etc.

Both the sacral and the coccygeal nerves also give off branches to the sympathetic ganglia in the pelvic area (see page 308).

For full details concerning the nerve supply of muscles and the distribution of the sensory nerves readers are referred to fuller textbooks on anatomy. The nurse does not normally require to know them.

The functions of the central nervous system. The functions of the central nervous system are:

(1) It is the seat of all *sensation*. Sensation is due to the bringing in of stimuli from the tissues by afferent fibres to the sensory centres of the brain. These stimuli pass through three or more afferent neurones before they reach the sensory centres of the cerebrum, by which they are interpreted as sensations. The meaning given to these sensations is due to experience. For example, the sound 'mother' has no meaning in the baby's brain when he first hears it, but by experience he learns gradually to associate with the word the sight of a certain face, the experience of being fed, bathed, changed and nursed, and the word 'mother' gradually comes to mean the person associated with all these experiences. Again, the word 'quack-quack' has no meaning, but if it is repeated to the young child every time he sees a certain picture, he learns to associate the word with a certain shape and colour. When the child goes into the park in his pram and sees the duck on the water, and the word 'quack-quack' is repeated, it comes to mean something which eats bread, swims, flies and walks.

(2) It controls all *movements* of *voluntary muscle*, i.e. the muscles of the head, limbs and trunk. Movement is here entirely due to nerve stimulus, and this can be set up in two different ways, so that movement is classified as: (a) voluntary movement, and (b) reflex movement. These are discussed in detail below.

(3) The brain is the seat of all the *special senses*, e.g. sight, hearing, smell, touch and taste.

(4) The cerebrum is the seat of all the *higher mental powers*—reasoning, will power, consciousness, memory, emotion, etc.

(5) The brain also controls the vital functions of *respiration* and *circulation*, the controlling centres being located in the hypothalamus and the medulla.

Voluntary movement is the result of stimuli produced at will in the motor centres of the brain. The stimuli are passed out by motor fibres through the cord and nerves to the muscles and produce contraction.

Reflex action is the result of the stimulation of the motor cells by stimuli brought in by afferent neurones from the tissues. Incoming stimuli can therefore, in addition to causing sensation, give rise to action. They only produce sensation if they are passed on to the sensory centres of the brain. On the other hand, in cord and brain they may stimulate the motor cells and give rise to action, *reflex action*. Sensory stimuli are pouring into the cord and brain from the tissues all the time. If they reach the sensory centres of the cortex of the cerebrum and stimulate them they produce sensations of which we are conscious. If they stimulate motor cells they produce reflex action, e.g. the touch of something hot on the skin causes immediate withdrawal of the part, a tap on the patellar ligament causes contraction of the quadriceps extensor muscle and produces the 'knee-jerk', stroking of the sole of the foot causes the toes to be drawn down. In the case of the first action it may be claimed that because the heat causes an unpleasant sensation we want to withdraw the part and the action is voluntary. On the other hand, the action occurs in an animal when the spinal cord has been severed in the neck, so that no stimuli can reach the brain, and is, in fact, more marked than in the ordinary animal. The sensory stimulus is brought into the cord by sensory fibres, is transmitted by connector fibres to the motor cells of the anterior horn, and is passed out by the motor fibres to the muscles. In the case of the knee-jerk the sensory stimulus is carried into the lumbar region of the cord by afferent fibres and stimulates the motor cells, which control the quadriceps.

The reason for the reflex being more marked when the cord is cut off from the brain is that the cerebral centres have an *inhibiting effect* on reflex action. Reflex action is not produced by the will; it is a mere response to environment, and is the only type of action found in the lower forms of life. With the development of the cerebrum, control of reflex action and the partial replacement of reflex action by voluntary action takes place. If we touch something hot the natural reflex is to withdraw, as the inexperienced puppy does when it picks up a hot cinder in the grate. On the other hand, if we pick up a hot plate with our dinner on it, appreciating the value of the plate and the dinner, we can continue to hold the plate and put it safely down at the expense of burnt fingers. The reflex is inhibited. If we have been warned and expect the plate to be hot, this control is more easily established. In the case of the pure reflex, where the action which is produced has no value, as in the knee-jerk and the drawing down of the toes on tickling the sole of the foot, there is still some inhibiting effect from the brain, and if the nerve path from brain to muscle is destroyed by injury or disease above the anterior horn cell the action is much more marked.

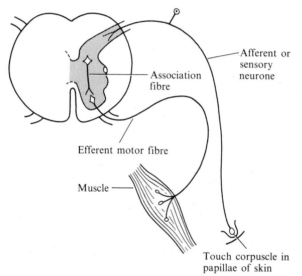

Afferent or sensory neurone

Association fibre

Efferent motor fibre

Muscle

Touch corpuscle in papillae of skin

FIG. 187. A reflex arc.

Actions which are in the first place voluntary become in a sense reflex, e.g. standing is in the first place a voluntary act which is carried out by the exercise of the will. When we have learnt to keep our balance on two feet we learn to do it by the sensations from the skin of our feet, and from muscles and joints, and the sensory organs of balance, and we can stand without voluntary effort unless disease affects the sensory nerves. In the same way, when we learn to knit it is at first a difficult voluntary action. Gradually our fingers and arms learn the feel of the wool, needles and actions, and we can knit without attention unless we lose the sensation in the parts or are learning a complicated new pattern.

Reflex actions may occur at three different levels in the nervous system:

(1) The spinal reflex, e.g. the knee-jerk.
(2) Reflexes occurring at the base of the brain, e.g. sneezing, coughing, vomiting, walking (cerebellar).
(3) Reflexes occurring in the cerebrum, and involving use of the association fibres of the brain.

Autonomic Nervous System

The autonomic nervous system supplies nerves to all the internal organs of the body and the blood vessels. It is so named because these organs are self-controlled (auto = self), and not under the control of the will. The functioning of the internal organs normally takes place without any conscious effort and without conscious knowledge. The will does not normally affect them, but the emotions do. They are affected by the hypothalamus.

The autonomic nervous system is for the most part efferent. It is made up largely of efferent neurones, both motor, supplying the involuntary muscles of the walls of organs such as the stomach, intestines, bladder, heart and blood vessels, and secretory, supplying the glands such as the liver, pancreas and kidney. There are some afferent fibres, but they are comparatively few in number, as the internal organs are almost insensitive. As a result disease may attack and destroy them without causing pain, and the pain which does occur is due to a large extent to inflammation of the lining membrane of the cavity in which they lie; for example, tuberculosis or pneumonia may affect the lung tissue without any pain, but as soon as the pleura is involved sharp pain is felt. In the same way it hurts to cut through the abdominal wall, but when a piece of bowel has been brought out of the abdomen it can be cut without causing the patient any pain, as the nurse may see in cases of colostomy.

The autonomic nervous system consists of two parts:

(1) The *sympathetic* nervous system.
(2) The *parasympathetic* nervous system.

The *sympathetic* nervous system consists of a *double chain of ganglia* running down the trunk just in front of the vertebral column in the cervical, thoracic, and lumbar regions. These ganglia are linked to one another by nerves. They receive nerves from the thoracic and upper lumbar regions of the

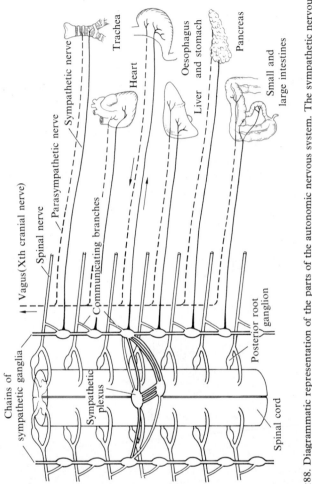

FIG. 188. Diagrammatic representation of the parts of the autonomic nervous system. The sympathetic nervous system is shown by a continuous line, and the parasympathetic nervous system by broken lines.

spinal cord. They give off; (a) nerves supplying the internal organs, *the visceral branches*; and (b) nerves running back to the spinal nerves, the *parietal branches*. These nerves supply the blood vessels, the sweat and sebaceous glands and the muscles which raise the hairs of the skin making them 'stand

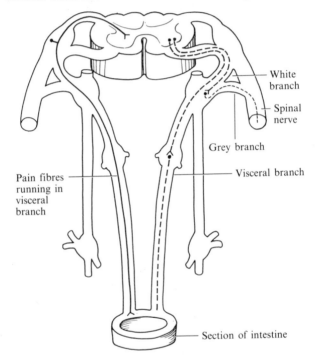

White branch

Spinal nerve

Grey branch

Visceral branch

Pain fibres running in visceral branch

Section of intestine

FIG. 189. Section of the spinal cord showing efferent and afferent pathways of the sympathetic nerves and the link with the spinal nerve. (After Cockett.)

on end'. In certain regions where there are many organs requiring a nerve supply there are additional ganglia between the two chains, linked up by nerves with the chains and with one another, and giving off nerves to the neighbouring organs; these are called plexuses, e.g. the cardiac plexus lies behind the heart in the thoracic region, and the solar plexus lies just below the diaphragm, where stomach, liver, kidneys, spleen and pancreas are all found.

The *parasympathetic* nervous system consists chiefly of the vagus nerve, which gives off branches to all the organs of the thorax and abdomen, but does also include branches from other cranial nerves (the third, seventh and ninth), and nerves from ganglia in the sacral region of the vertebral column.

The Functions of the Autonomic System

All the internal organs have, therefore, a double nerve supply, sympathetic and parasympathetic, and the two sets of nerves have in each case opposite actions, one *stimulating* and the other *checking* the activity of the organ. The arrangement is similar to a motor-car, with an accelerator pedal to make it move faster and a brake pedal to check it. The *sympathetic* nerves have a stimulating and quickening effect on the heart and on the respiratory system, but a checking effect on digestion. They improve the circulation and cause dilatation of the bronchial tubes, increasing the air intake, but they stop the secretion of digestive juices from the salivary glands, and throughout the alimentary canal, and check peristaltic action in its wall. These nerves are stimulated by strong emotion, such as fear, anger and excitement. (It is because of this effect of the emotions that they are called sympathetic.) Their functions are thus closely linked to those of the adrenal medulla, which they stimulate. They help to enable the body to respond to emotion, since they provide the muscles with a better supply of blood which is rich in oxygen. This enables the individual to run away when frightened or to fight when angry, the instinctive response to these emotions. On the other hand, they are responsible for the arrest of the digestion of food in strong emotion, and thus may produce vomiting and emptying of the bowel, the organs getting rid of the contents which they cannot cope with for the time being.

The parasympathetic nerves have exactly opposite effects, stimulating the digestive system and producing both a copious flow of digestive juices and peristaltic action. On the other hand, the vagus slows the heart, reducing the circulation, and has a checking effect on the respiratory system, contracting the bronchial tubes. These nerves are stimulated by pleasant emotions. As a result, happiness and a contented mind tend to improve digestion. Pavlov was able to show this on a dog with a gastric fistula. When the dog was shown a bone which pleased him, gastric juice began to pour out into the stomach by a reflex action through the vagus nerve. The bringing of a cat into the room angered the dog and the flow was checked, the sympathetic nerves being stimulated. Pavlov also noticed that if a bell were rung before the dog's food was brought in each day, after a time the ringing of the bell without the bringing in of food would cause a flow of gastric juice. This is termed a *conditioned reflex*. The animal had learned to associate the two things, so that either would produce the same reflex response. It is the same with human beings. The ringing of a dinner bell will cause secretion of digestive juices, just as will the sensation of the smell and sight of pleasant, well-served food. In the invalid, whose digestion is impaired, it is therefore of special importance to serve foods appetizingly, and choose dishes which will give the patient as much pleasure as possible.

24 The Ear

THE EAR is the *organ of hearing*. It is designed to collect sound waves so that they stimulate the endings of the nerve of hearing. The nerve transmits the stimulus to the centre of hearing in the brain, by which it is appreciated as sound and interpreted because of previous experience of similar sounds.

The *sound wave* is a wave of *vibration* of the air set up by the vibration of an object. The vibration of a fiddle-string or the vocal cord, for instance, sets up vibration of the air in contact with it and produces waves of vibration which spread out in all directions, like the ripples set up in a still pond when a pebble is thrown into it. The sound waves gradually get less as they travel from the vibrating object and ultimately die away. If they hit a wall at right angles to their course they are reflected back and produce an echo. It is this reflection of sound waves which affects the acoustics of any hall, making it either easy or difficult to speak so as to be heard throughout it.

To produce sound, vibrations must be within certain rates. The human ear is stimulated only by vibrations at rates between 30 and 30,000 per second. The slow vibrations produce low notes and the quick vibrations produce high ones. It is for this reason that a man's voice is lower than a woman's, for his vocal cords are longer and vibrate more slowly, while the woman's are shorter and vibrate more quickly. It is the rapid growth in the larynx and the consequent lengthening of the boy's vocal cords at puberty which causes the breaking of the voice. It is probable that the ears of animals respond to a different range of vibrations, so that they can hear sound when we can hear nothing.

Sound waves travel at the rate of 1090 ft per second. They travel much more slowly than light rays, hence the flash of lightning is seen before thunder is heard, and the farther away the storm the longer the interval between the two.

Sound waves normally are carried by air, but they also pass through solid bodies; in fact, a solid carries sound more readily than air. Thus, putting an ear to the ground, it is possible to hear footsteps at a greater distance than through the air. It is, however normally air with which the ear is in contact.

The ear consists of three parts:

(1) The *external* ear.
(2) The *middle* ear.
(3) The *internal* ear.

The *external ear* consists of:

(1) The auricle.
(2) The external acoustic meatus.

The *auricle* is the part which stands out from the head. It consists of a piece of elastic fibrocartilage folded round, and leading to, the external acoustic meatus like a funnel. It is covered with skin and richly supplied with blood vessels to keep it warm. These are necessary because of its exposed position, and as a result of the good blood supply injuries to the auricle readily heal. In fact, if it is completely cut off, it can be stitched on again and will remain alive until a new blood supply has developed to link it to the circulation again.

The auricle is not of much functional value in man, though it serves to some extent as a funnel collecting up the sound waves and carrying them into the acoustic meatus. In animals it often has two other functions. It is protective, hanging down over the meatus and preventing twigs or thorns from causing injury to the drum, and also it is freely movable and can be directed towards the point from which sound waves are coming, so that it collects them better and improves hearing. You will see that a horse's auricles, unless the animal is deaf, are constantly in movement as it passes along a street, as are the human eyes. Actually the human auricle is provided with muscles for its movement, but these, from disuse, are now very little developed, though some comedians cultivate their use with amusing results.

The *external acoustic meatus* is a tunnel about 4 cm long leading into the temporal bone. The outer part has walls of cartilage, the deep part bony walls. Its far end is completely blocked by a sheet of fibrous tissue called the *tympanic membrane*. The canal is lined throughout with skin, which covers the outer surface of the drum. The canal is curved; it runs first upwards and backwards, then downwards and forwards. These curves, together with the depth of the canal, serve to protect the tympanic membrane, which is essential for hearing. These curves make the cleansing of the canal by syringing difficult, but it can be straightened by pulling the outer cartilaginous part into line with the rigid deep part. In the adult an upward and backward pull on the auricle does this, but in children a backward pull, and in infants a pull downwards, backwards, and outwards is necessary, since ossification is not yet complete.

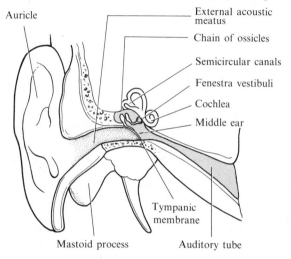

FIG. 190. The ear.

The skin at the entrance to the canal carries *hairs* and *glands* which secrete wax, known as *cerumen*. Both these protect the canal from the entrance of foreign bodies, the hairs acting as a filter and the wax entangling dust and insects. Normally the wax tends to dry up and fall out with the movements of the jaw as we chew our food. If, however, it is a little changed in character, and food is soft and requires little chewing, the wax tends to collect and may block so much of the canal that a little water getting in during washing will complete the obstruction, and no sound waves can pass along. This causes complete deafness in that ear, with buzzing noises in it. It is easily remedied by syringing out the canal after softening the wax, if necessary.

The *middle ear* is a tiny cavity in the temporal bone between the external and internal ears. The tympanic membrane separates it from the external ear, and two little windows in its wall, the fenestra vestibuli and the fenestra cochlea, lead from it to the internal ear. The last of the ossicles fills the fenestra vestibuli and the fenestra cochlea is filled in with fibrous tissue. The cavity contains air, which enters it by the auditory tube from the nasopharynx. This air is essential to allow the drum to vibrate freely; if the tube is blocked and no air can enter, that in the cavity is gradually absorbed, and the air pressure on the outer surface becomes the greater, driving the drum in and stretching it and so stopping its vibration. The middle ear is lined with mucous membrane continuous with the lining of the auditory tube. Hence infection from the nasopharynx in a cold or in measles, scarlet fever, or diphtheria, may easily spread up the tube and infect the middle ear, causing inflammation there.

A *chain* of three tiny bones, or *ossicles*, linked to one another by ligaments, runs across the middle ear from the drum to the fenestra vestibuli. They are called the *malleus*, the *incus*, and the *stapes*. These bones carry vibration across the middle ear to the internal ear. Sound waves entering the middle ear set the tympanic membrane vibrating; this makes the ossicles vibrate and sets the fenestra vestibuli vibrating.

The middle ear cavity also opens into a cavity in the mastoid process known as the mastoid antrum, which in turn leads to a number of other smaller cavities called the mastoid air cells. These all contain air, which enters via the auditory tube and middle ear, and they are lined with mucous membrane continuous with the lining of the middle ear. As a result, infection from the throat and ear may spread to the mastoid process and cause pus to form there. This is very serious, as the bony roof of the mastoid antrum is thinner than the outer wall, so that pus can more easily escape upwards into the cranial cavity than outwards. The surgeon therefore operates and allows the pus to escape unless early treatment of the infection clears up the condition. Otherwise the brain or meninges might become the seat of septic infection.

The *internal ear* is a complicated cavity in the petrous portion of the temporal bone at the base of the cranium. This complicated cavity is known as the bony *labyrinth*. It consists of three chief parts:

(1) The *vestibule*, a small cavity behind the fenestra vestibuli, to and from which the others lead.

(2) The *cochlea*, which is shaped like a tiny snail's shell; this is the part through which we hear.

(3) Three fine *semicircular canals* leading out from and back into the vestibule in the three different planes of space, one running up and down, one sideways, and the other backwards and forwards; these give sense of position.

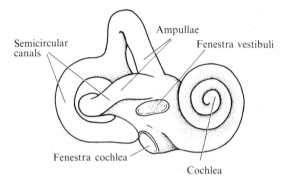

FIG. 191. The bony labyrinth.

The *cochlea* is a spiral tube which makes two and three-quarter turns round a central pillar of bone called the modiolus. The tube is divided lengthwise into three separate tunnels by two membranes, the basilar membrane and the vestibular membrane, which stretch from the modiolus to the outer wall. The outer tunnels are the scala vestibuli above and the scala tympani below. These tunnels are filled with perilymph and they join at the top of the modiolus. The lower end of the scala tympani is closed by the fibrous fenestra cochlea. The middle tunnel is called the cochlear duct and is filled with endolymph. It is the same shape as the bony labyrinth and is called the membranous labyrinth. Within the cochlear duct are the special nerve endings of the auditory nerve called hair cells. Hearing is due to the sound waves setting the tympanic membrane in vibration; this sets the ossicles and fenestra vestibuli vibrating, which in turn causes vibration of the perilymph. As fluid is incompressible, the perilymph can vibrate only if the fenestra cochlea is able to bulge outwards as the fenestra vestibuli bulges inwards. Hence the need for two windows in the internal ear. Vibration of the perilymph gives rise to vibration of the endolymph, and this affects the little hairs which jut into it and stimulates the endings of the vestibulo-cochlear nerve in the membranous cochlea. The nerve carries the stimulus to the centre of hearing in the temporal lobe of the brain, where it is appreciated and interpreted.

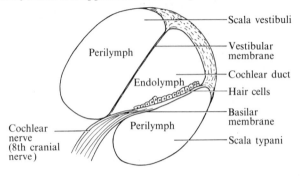

FIG. 192. Section through the cochlea.

The appreciation of sound will result from any stimulus brought by the auditory nerve to the centre of hearing, but the meaning given to the sound will depend on previous experience and the power of reasoning. Thus the word 'mother' has no meaning to a baby at first, but comes to be associated with the woman he sees and is aware of in other ways through his various senses when the sound is uttered, and so it gains meaning by association. If someone told you about the ear in Chinese, you would know from previous experience that you

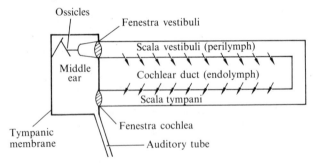

FIG. 193. Diagram to show how vibrations pass through the inner ear.

were listening to speech, but you would know nothing more about the ear unless you had learnt Chinese, as the sounds would have no association in your brain.

By the *semicircular canals* we can appreciate *position* in space. When we change our position the fluid in the semicircular canals moves and exerts pressure on the nerve endings there. Thus with eyes bandaged we could tell if anyone altered our position. This power to appreciate position is essential for the keeping of balance. As we know our position we are able to maintain it through our muscles. Over-stimulation of the canals causes giddiness; we do not any longer know where we are, and therefore cannot maintain the position, with the result that we are liable to stagger and reel or, in extreme cases, fall to the ground. Giddiness is commonly associated with ear disease, since these two sense organs are combined in the internal ear.

25 The Eye

THE EYE is the *organ of sight*. It admits light rays and enables them to stimulate the endings of the optic nerve. The nerve carries the stimulus to the centre of vision in the brain, causing sight.

Light rays are electro-magnetic waves, unlike the sound waves that are waves of vibration in the air. The light ray is similar to the heat ray, ultraviolet ray, X-ray and to the wireless waves, but differs from them in wavelength and vibration rate. Heat rays are waves of greater length and slower vibration than the light ray, while ultraviolet rays, X-rays, and radium rays are shorter and vibrate more rapidly. All these waves are very minute, their length being only a tiny fraction of a millimetre, except for the waves used in wireless telephony. Some wireless waves measure many metres in length. All these waves travel very rapidly through space. They travel at 186 000 miles per second. Hence messages can be flashed round the Earth's surface in a fraction of a minute.

A ray of light can be split up by passing it through a glass prism into all the colours of the rainbow. The various colours are produced by waves of varying length; the red colour on the one side of the rainbow has the longest wavelength and the violet on the opposite side the shortest. When light rays strike an object they may either pass through it, in which case the object is said to be transparent, or may be reflected or absorbed by the object, in which case the object is said to be opaque. If the light ray is reflected completely the object appears white. If the ray is completely absorbed the object appears black. If the ray is partly absorbed and partly reflected the object appears to have the colour of the reflected rays, e.g. if red rays are reflected, but all the other rays are absorbed, the object appears red.

It is through these light rays alone that we are able to see. In a dark room well-known objects may surround you, but with no light rays they have neither colour nor form, and the eyes are useless for sight.

The *eye* is a ball or sphere about 2·5 cm in diameter. It lies in the bony orbit, resting on a pad of fat and attached to the orbit by muscles. There are *six muscles* to each eye, four straight muscles or *recti* (the superior, inferior, lateral and medial recti) and two *oblique* or sloping muscles (the superior and inferior obliques). These muscles can move the eye freely in all directions, up and down, in and out.

The recti run to the eyeball from the back of the orbit. They arise from the bony orbit around the opening through which the optic nerve passes. This opening is not in the centre of the back of the orbit, but towards the nasal side in each case. The superior and inferior recti therefore do not pull the eyeball straight up and down, and the oblique muscles are used with them, the superior oblique with the inferior rectus and the inferior oblique with the superior rectus, to correct the pull towards the midline and provide for straight up and down movements of the eyeball.

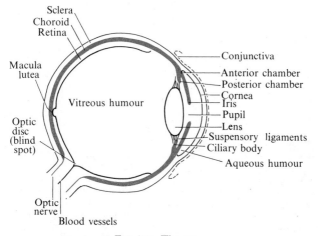

Fig. 194. The eye.

The *optic nerve* joins the eyeball a little to the nasal side of the centre of the back of the eye. It is somewhat like the stalk of a cherry, but much thicker in comparison.

The eyeball has a *wall* made up of three coats:

(1) The *sclera* and *cornea*, a fibrous protective coat.

(2) The *choroid*, *iris* and *ciliary body*, a pigmented, vascular coat.

(3) The *retina*, a nervous coat, containing the nerve endings of the optic nerve.

The *sclera* is a tough coat of fibrous tissue which forms the *white* of the eye. It is strong and protective in function. It does not allow light rays to pass through it.

The *cornea* is a circular, transparent area in the front of the eyeball continuous with the sclera. It acts as a *window*, admitting light rays into the eye. It covers the coloured part of the eye and projects slightly like a watch-glass.

The *choroid* is a pigmented coat which lines the sclera, forming a dark lining which absorbs the light rays, preventing their reflection within the eye. This is essential for sight. The

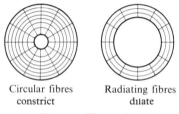

Circular fibres   Radiating fibres
constrict   dilate

Fig. 195. The pupil.

iris is the coloured part of the eye, blue, brown or grey. It lies behind the cornea, and has a circular opening in the centre, the *pupil* of the eye. This looks black, as all the light which enters the eye is absorbed by the choroid. The pupil is thus merely the hole in the iris by which light rays enter the eye. The *iris* acts as a shutter controlling the amount of light which enters. It is muscular, and contains both circular and radiating muscular fibres. The circular fibres contract the pupil and are supplied by the third cranial nerve. The radiating fibres dilate the pupil and are supplied by sympathetic nerves. The pupil is contracted in a strong light to prevent too much light entering and over-stimulating the nerve endings, for too much light is blinding. It is also contracted for near sight. The pupil is dilated in a poor light and for far sight.

Where the choroid and iris meet, round the border of the cornea, there is a thickening of this pigmented coat called the *ciliary body*. This contains both muscle and glands. The *ciliary muscle* controls the lens and is called the muscle of *accommodation*, as by it the lens is focused for near or far sight as required. The *ciliary glands* produce a watery fluid, the aqueous humour, which nourishes the internal parts of the eye. This is necessary, as the contents of the eyeball have no blood supply, since blood is opaque and would stop the passage of light rays to the retina (see also p. 323 for action of ciliary muscle).

The *retina* is the nervous coat. It is composed of the *fibres* of the optic nerve and the *nerve endings*. It is very complicated and has many layers, of which the inner and outer are most important to the nurse. The inner layer consists of the fibres of the optic nerve which passes through the coats of the eyeball from behind and spreads out its fibres to line the eyeball, rather like a wine-glass with the optic nerve as the stem of the glass. The fibres end by dipping into the retina and linking up with the *rods* and *cones* of the outer layer. The rods and cones are the specialized nerve endings sensitive to light; they form the essential part of the layer of the retina which is in contact with the choroid. The rods and cones are thought to have different functions, the cones being sensitive to colour and bright light, and the rods to dim light. They are not distributed evenly over the retina, the cones being more numerous towards the centre of the back of the eye and the rods in the outer part, which is used for sight in a dim light. At a point immediately opposite the centre of the cornea, the cones are particularly close together, forming what is known as the *macula lutea*; here sight is most perfect. From this point they get gradually less and less numerous as the retina passes outwards and forwards, till they finally disappear altogether in the front of the eyeball, since no light rays can strike there, as there is no reflection within the ball and all light enters from the front. The rods and cones are in contact with a layer of pigment cells, and the rods contain a pigment known as the visual purple. Light rays which strike the retina pass through it and are absorbed by the dark choroid below. As they pass, they produce a chemical change; they bleach the visual purple in this chemical change, and this stimulates the nerve endings, producing an impulse which is carried by the fibres of the optic nerve to the centre for vision in the occipital lobe of the brain, where it is interpreted as sight. To see anything clearly and in detail it must be focused on the macula lutea of the retina. Thus to tell the time by a clock you must look directly at it, but you can see it and recognize that it is a clock when its image is focused on some other part of the retina, though you cannot read its figures nor see where its hands lie.

After the visual purple has been bleached by light it has to be re-formed, and for this purpose vitamin A from the blood stream is needed. Vitamin A and a protein form the visual purple, so that if there is a deficiency of vitamin A in the diet visual purple is not produced in the normal quantities and *night blindness* results, as the rods are responsible for sight when light is poor. Because of the function of the rods for sight when the light rays are weak, it is often possible to see faint stars if you do not look directly at them, so that their light strikes the outer part of the retina where the rods are most numerous. If the eyes are turned to look directly at these lesser stars, they disappear, as their light rays are not bright enough to stimulate the cones.

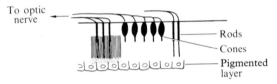

FIG. 196. Layer of rods and cones on the outside of the retina.

Where the optic nerve enters there are no nerve endings, and therefore there is a *blind spot* in each eye, called the optic disc, with which we can see nothing. This is not obvious when we use both eyes, as an object can only be opposite the blind spot in one eye at a time. To find the blind spot, mark a piece of paper as shown below:

X   ●

Now, shut the left eye, fix the right eye steadily on the cross, and move the paper slowly backwards and forwards on eye level. At a certain point the dot will disappear because it is opposite the blind spot.

The eye contains:

(1) The lens
(2) The aqueous humour
(3) The vitreous humour

The *lens* is a biconvex crystalline body lying in the front of the eyeball immediately behind the iris. It is enclosed in a transparent capsule, from which ligaments run out in all directions to the ciliary body; these ligaments are called the *suspensory ligaments*, since they hold the lens in position. The lens

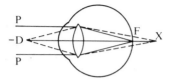

FIG. 197. The normal eye at rest. P = Parallel rays from a distant object focused by the lens on the retina (F). Dotted lines show the diverging rays from a near object (D) from which the focus falls behind the retina at X.

of the eye is a highly elastic body, unlike the glass lenses we use as magnifying glasses. Normally when the eye is at rest the suspensory ligaments exert a slight pull on it, flattening it and lessening its curvature. The lens is the means by which the focus of the eye is altered for far or near sight. The focusing power of the lens depends on its curvature. The greater the curvature of its surfaces, the stronger is its focusing power. A stronger lens is required for near sight than for far sight. Focusing is the bending of the light rays given off by an object so as to produce an exact image of the object from which the rays come. The rays from a far object which hit and enter the eyeball are almost parallel to one another; those from a near object are diverging from one another. The rays from a near object therefore need to be more sharply bent than those from a far object to bring them to focus on the retina. For *far sight* the resting eye is suitable, as the lens is flattened and does not bend the light rays much. For *near sight* the ciliary muscle in the ciliary body contracts. This relaxes the pull which the suspensory ligaments exert on the lens from all directions. When the pull is relaxed the elastic lens thickens and becomes more curved, and can focus the near object on the retina.

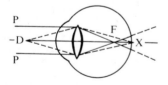

FIG. 198. The normal eye during accommodation. Dotted lines from a near object (D) are focused on the retina at point X by the more curved lens. P = Parallel rays from a distant object which are now focused in front of the retina (F).

The actual picture produced on the retina is upside down as a result of the bending of the light rays by the lens. Since the power of sight is the interpretation of the stimuli by the brain we see the object the right way up. The bending of the light rays by the lens is called refraction. Refraction of light rays occurs when light rays pass from one transparent medium to another of greater density. All the contents of the eyeball, the lens, and the aqueous and vitreous humour, help in the refraction of light rays, since they are denser than the air, but the lens is the one which can be altered to accommodate the sight for near and far vision.

FIG. 199. A convex lens, showing how parallel light rays are brought to focus at F.

Remember that near vision means muscular work, while far vision is provided for when the eye is resting. Near work, such as sewing or reading, is therefore a strain on the eyes, and this should be relieved by lifting the eyes from near work at frequent intervals.

Abnormal Sight

Some people are naturally *short-sighted*. This is due to the fact that their eyes are too long, so that the retina is farther from the lens than it should be and the focusing point lies in front of it. In this case the near object can be seen with the flatter lens of the eye at rest (normally used for far sight), but for far sight concave glasses are necessary to throw the focus point farther back.

Others are naturally *long-sighted*. This is due to the fact that their eyes are too short, so that the retina is too close to the lens and the focusing point lies behind it. In this case the far object can be seen by using the thicker, more curved lens used by the normal person for near sight, as this will bend the light rays more and bring the focusing point forward; for near sight, since they are already using the more curved lens, they must be provided with convex glasses to bend the light rays from the near object still more acutely.

The *aqueous humour* is a watery fluid, which fills the eyeball from the lens forward to the cornea. It nourishes the internal structures of the eye, which have no blood supply, and acts as a refracting medium, bending the light rays and helping to focus them on the retina. It is constantly secreted by the ciliary glands and absorbed into the circulation.

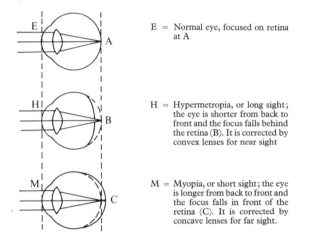

E = Normal eye, focused on retina at A

H = Hypermetropia, or long sight; the eye is shorter from back to front and the focus falls behind the retina (B). It is corrected by convex lenses for near sight

M = Myopia, or short sight; the eye is longer from back to front and the focus falls in front of the retina (C). It is corrected by concave lenses for far sight.

FIG. 200. The normal eye, hypermetropia and myopia.

That part of the eyeball which is filled by the aqueous humour is divided anatomically into the *anterior chamber* and the *posterior chamber*. The anterior chamber is the space between the iris and the cornea. The posterior chamber is the space between the iris and the lens and suspensory ligament. The two communicate with one another through the pupil, which is the opening in the centre of the iris.

The *vitreous humour* is a jelly-like material which fills the eyeball from the lens backward to the retina. It serves as a refracting medium, bending the light rays as they pass through it, and helping in focusing, and also keeps the retina in position. If it escapes from a wound in the eye it cannot be renewed and the retina falls forward out of the correct position on which the light rays are focused and sight is lost.

The visual path. The optic nerve fibres run from their receptor endings in the retina to the optic disc, where they pass out of the eyeball, forming the optic nerves. These run back and towards the midline for a short distance to meet one another at the optic chiasma, which is just in front of the stalk of the pituitary gland.* Here the fibres of each nerve divide, those from the nasal side of each eye crossing over and joining those from the temporal side of the other eye to form the optic tracts, right and left, which run back to the lateral geniculate bodies below the thalami. From here other neurones transmit the impulses to the visual centres in the occipital lobes of the two hemispheres. Division of the optic nerve thus causes blindness

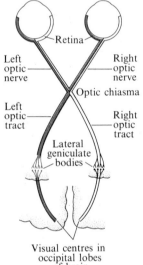

Retina

Left optic nerve

Right optic nerve

Optic chiasma

Left optic tract

Right optic tract

Lateral geniculate bodies

Visual centres in occipital lobes of brain

FIG. 201. The visual path from the eyes to the visual centres.

of the eye on that side, but division of the optic tract causes half-blindness in each eye, as no stimuli can reach the brain from the outer side of the eye on the same side of the head and the inner side of the eye on the opposite side of the head. This reduces the field of vision considerably. Man has *binocular* or *stereoscopic* vision, i.e. his eyes are so placed that he can see

* Note. Hence tumours of, or operation for removal of, the pituitary gland may result in damage to the optic nerve fibres either in front of the optic chiasma or at this point.

objects with both eyes at the same time, whereas many mammals have their eyes at the side of the head, so that they cannot direct them both towards any one object. Man learns to fuse what he sees with each eye into one picture and gains the power to appreciate the distance of an object from his eye and its thickness. This appreciation of depth and distance is termed stereoscopic vision. By closing one eye and using one only, it is easy to realize the flatness which results from the lack of binocular vision.

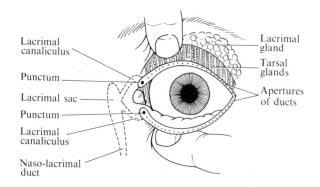

FIG. 202. The lacrimal apparatus.

The Protection of the Eyes

The eyes are very delicate organs and are protected by the eyebrows, the eyelids, and the lacrimal apparatus, as well as by the bony orbits in which they lie embedded in fatty tissue.

The overhanging *brows* protect them from injury and excessive light, while the hairs entangle sweat and prevent it from running into the eyes.

The *eyelids* consist of a plate of fibrous tissue covered by skin and lined with mucous membrane. The edges of the lids are provided with hairs, the eyelashes, which keep out dust, insects, and too much light. The transparent mucous membrane which lines the lids is reflected over the front of the eyeball and is called the *conjunctiva*. This results in the formation of upper and lower conjunctival sacs under the upper and lower lids respectively. Dust and bacteria tend to stick to the moist surface of this membrane, and to keep it clean it is constantly washed by the lacrimal apparatus.

The *lacrimal apparatus* consists of:

(1) The *lacrimal gland*, lying over the eye at the outer side and secreting lacrimal fluid into the conjunctival sac.

(2) Two fine canals, called *lacrimal canaliculi*, leading from the inner angle of the lids to

(3) The *lacrimal sac*, which lies at the inner angle of the eyelids in the groove on the lacrimal bone.

(4) The *naso-lacrimal duct*, which runs from the lacrimal sac down to the nose.

The opening into the canaliculi can be seen at the inner angle of the eyelids and is called the *punctum* (plural, *puncta*).

The fluid secreted by the lacrimal glands washes over the eyeball and is swept up by the blinking action of the eyelids. The muscles which cause the blink press on the lacrimal sac and contract it, so that as they relax the sac expands and sucks the fluid from the edges of the lids along the fine canals into the sac; from this it runs by gravity down into the nose. Thus the window which admits light into the eye is constantly irrigated by a gentle stream of fluid, which keeps it clean and washes away germs and harmful substances. The fluid is composed of water, salts and an antibacterial substance called lysozyme.

26 The Skin

EXCRETION is the discharge of waste products. Everything which the body takes in and uses gives rise to waste products which must be excreted if health is to be maintained. The use of proteins gives rise to the formation of urea, uric acid and creatinine; the use of fuel foods to the formation of carbon dioxide and water. The water and salts taken in produce water and salts for excretion.

The *excretory organs* include:

(1) The *skin*, excreting sweat.
(2) The *kidneys*, excreting urine.
(3) The *lungs*, excreting carbon dioxide and water.
(4) The *liver*, excreting bile.

The bowel or large intestine expels the faeces which contains ingredients derived from the bile, including bile pigments which colour the faeces. It is not, strictly speaking, an excretory organ in the same sense as the other organs in this list, since faeces consists mainly of material which has merely passed through the body, i.e. cellulose and bacteria, and which has not been picked out of the blood for excretion. On the other hand, it contains water, salts, and bile pigments which have been excreted by the liver and to a lesser extent by the mucous membrane lining the canal.

The sweat consists of water and salts. The urine contains water, salts, urea and the other waste products of protein metabolism. The expired air contains the waste products of combustion, carbon dioxide and water. The bile contains water, salts and bile pigments.

Most of the organs included in the list of excretory organs have other functions, and have therefore been described already under other systems. Only the skin has not been dealt with yet.

The Skin

The skin covers the entire surface of the body. It is not, however, merely a covering like the brown paper on the outside of a parcel. It is a vital organ essential to health and life. It consists of two coats:

(1) The *epidermis* or outer layer.
(2) The *corium* (dermis), the deeper layer.

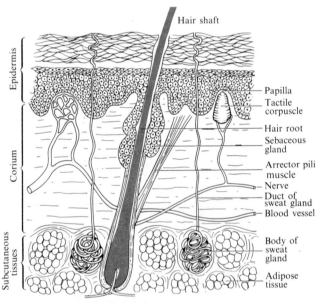

Hair shaft

Epidermis

Corium

Subcutaneous tissues

Papilla
Tactile corpuscle
Hair root
Sebaceous gland
Arrector pili muscle
Nerve
Duct of sweat gland
Blood vessel
Body of sweat gland
Adipose tissue

Papilla of the hair

FIG. 203. Diagrammatic section of the skin.

Beneath these is a layer of fatty tissue which attaches the skin loosely to the fascia of the under-lying muscle; this is known as the *subcutaneous tissue*.

The *epidermis* is composed of *stratified squamous epithelium*; it consists of many layers of cells, but varies greatly in thickness according to the friction to which it is exposed. It is thickest on the sole of the foot and the palm of the hand, where there may be sixty layers of cells or more. It is thin between the fingers, between the toes, and on the inner sides of the limbs and the abdomen.

The epidermis is divided into two layers:

(1) The horny zone.
(2) The germinative zone.

The horny zone has three layers:

(1) The horny layer (stratum corneum) is the most superficial layer. The cells are flat and have no nuclei and the protoplasm has been changed into a horny substance called keratin, which is waterproof.

(2) The clear layer (stratum lucidum) is composed of cells with clear protoplasm and some have flattened nuclei.

(3) The granular layer (stratum granulosum) is the deepest layer. It consists of several layers of cells with granular protoplasm and distinct nuclei.

The germinative zone, which is the deeper, consists of two layers:

(1) The prickle cell layer contains cells of varying shapes, each having short processes, rather like thorns, joining them together. The nuclei are distinct.

(2) The basal cell layer consists of columnar cells arranged on a basement membrane.

The surface scales are constantly being rubbed off by friction and are constantly renewed from below as the deep cells multiply and are driven up to the surface, developing into scales as they approach it. The epidermis has no blood supply and practically no nerve supply. It is nourished by lymph from the blood vessels in the underlying corium. It is the epidermis which is raised in a blister, and the blister can therefore be snipped with scissors without causing any pain, to allow the lymph it contains to escape.

The basal layer of cells in contact with the corium contains the pigments which give the skin its colour: white, yellow, red or black. Colour is therefore not even skin deep, being in the epidermis and not in the true skin. The pigment protects the body from the harmful effects of the sun's rays, since dark colours absorb radiation. The white skin of Europeans is due to the protection of the skin from sunlight by the wearing of clothing and living in houses. If the white skin is exposed to sun and air, as in sun-bathing, pigmentation results, but until this has occurred over-exposure will result in painful burns, with possible blistering. Some people have a particularly sensitive skin which freckles without becoming generally browned; they cannot tolerate exposure to heat so readily and must take special care to avoid over-exposure, as must the person whose skin is white from an indoor life.

The scales of the surface prevent the entrance of bacteria into the tissues, since they cannot digest these dried-up cells and make their way through them. Once the epidermis is broken by a cut or prick, infection may enter the tissues and sepsis may follow.

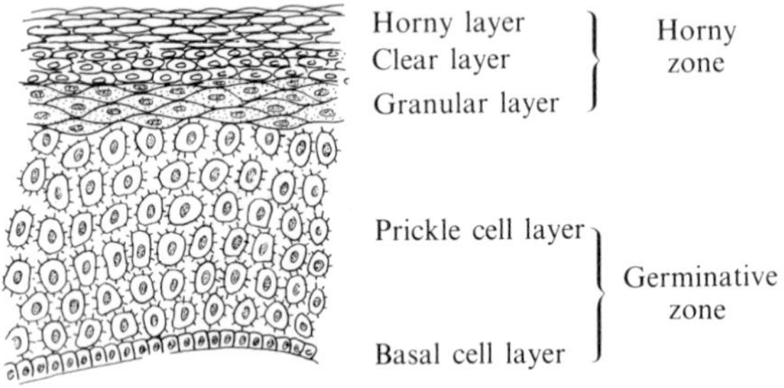

FIG. 204. The epidermis.

The corium is a tough elastic coat of *fibrous tissue*. It is this coat of an animal's skin which is tanned to make leather, and this shows how strong it is. It is very richly supplied with *blood vessels* and *nerves*. Its surface layer contains tiny cone-shaped projections or papillae, which jut up into the overlying epidermis. These increase the surface for the distribution of nerve-endings and blood vessels. They are most numerous where feeling is most acute, e.g. the tips of the fingers, toes, the palm of the hand, and the sole of the foot. It is the massing of the papillae in ridges here which gives the lined effect in these parts seen in the finger-print.

The *nerve-endings* in the skin are for the most part sensory, and are of different varieties to give the various different sensations of which the skin is capable—namely, the sensations of touch, heat, cold and pain. The nerves of touch end in round bodies known as the touch or tactile corpuscles which are stimulated by pressure, and the nerves of heat, cold and pain in delicate, tree-like branches. A few of the branches of these nerve-endings pass into the epidermis. Heat is experienced only if a hot thing touches the skin over the ending of a special nerve-ending affected by heat. In some parts the nerve-endings are so close together that this is not appreciated, but where the nerve-endings are less numerous, as over the back of the hand, it is possible to find spots where heat can be felt (hot spots), others where cold is felt (cold spots), and so on.

The *blood vessels* of the true skin are very numerous, not because the skin is active tissue, but because of its important function in cooling the body. The blood which circulates in the skin is exposed to the cool air which moves over the skin and loses heat to it by radiation and conduction. On the other hand, the blood in the muscles and deep organs is protected from heat loss by the layer of subcutaneous fat which lies below the skin. In hot weather and in exercise, when much heat is produced within the body, the blood vessels of the true skin are dilated by the vasomotor nerves which supply their walls. As a result, much blood flows through the skin and much heat is lost to the air. It is estimated to contain from one-third to even one-half of the blood when all its blood vessels are fully dilated. In cold weather and in rest the blood vessels of the true skin are contracted, so that comparatively little blood circulates through the skin and little heat is lost, the blood remaining in the deep organs, where it retains its heat. In this way the skin helps to regulate body temperature, preventing either over-heating or excessive cooling in normal circumstances.

The deeper layers of the true skin in contact with the subcutaneous tissue contain a certain amount of fat, so that the transition from one coat to the other is gradual.

Appendages of the skin. The skin carries the following appendages:

(1) Sweat glands
(2) Hairs
(3) Nails
(4) Sebaceous glands

The *sweat glands* are twisted, tubular glands which lie deep in the true skin. Their ducts open in the pores of the epidermis, and the tube is coiled up into a little round ball deep in the skin, called the body of the gland. The sweat glands secrete sweat, which consists of water, salts, and a trace of other waste products. Much of this sweat evaporates immediately it reaches the skin surface, and is called *insensible sweat*. When sweating is heavy some of the sweat is poured out on to the skin, which becomes wet with sweat; this is called *sensible sweat*, and evaporation takes place from the whole surface. If sweating is profuse and runs off the body the cooling effect is lost. The secretion of sweat is a means of excretion of waste products, and some poisons and drugs may be got rid of in this manner; its chief importance, however, lies in the fact that the evaporation of sweat uses up body heat, since heat is required to turn water into water vapour. The amount of sweat secreted therefore depends on the amount of heat that the body needs to lose. The average amount excreted in 24 hours is 500 to 600 ml. In hot weather and in violent exercise sweating is heavy, so that much heat is lost by the evaporation of sweat. At these times less urine is passed, so that there is not excessive loss of fluid. In cold weather or rest, less sweat is secreted, so that there is less loss of heat by evaporation; at the same time more urine is secreted by the kidney, so that loss of water is balanced.

Sweat glands are present all over the body, but are larger and more numerous in certain parts, such as the palm of the hand, the sole of the foot, the axillae, groins and forehead.

The *hairs* consists of modified epithelium. They grow from tiny pits in the skin, known as hair follicles. At the base of each follicle is a group of epithelial cells which forms the root from which the hair grows. The hair root is the part of the hair within the follicle, extending through both corium and epidermis. The hair shaft projects beyond the epidermis. The hair bulb is the expanded part of the hair within the follicle. At its base is a small conical projection called a papilla which contains blood vessels and nerves to supply the hair. Hairs are always set obliquely in the skin. The arrectores pilorum are small involuntary muscles connected to the hair follicles. They are always on the side toward which the hair leans so that when they contract the hair stands erect. The skin around the hair is elevated at the same time, which causes the appearance known as 'goose flesh'. The hairs are constantly being shed and renewed. As long as the root is healthy, fresh hair will grow from it, but if the root is destroyed or its blood supply interfered with, the growth of hair will stop. On the scalp baldness results. Good brushing, which stimulates the blood supply, and massaging of the scalp promotes the health and growth of the hair of the head. Hair is present all over the body except on the palms of the hands and the soles of the feet, but is so fine and sparse as a rule that it cannot be seen—a point the nurse should remember in preparing the skin for operation. It is longer and more plentiful in the eyebrows, axillae, and groins, where it entangles the sweat, preventing it from running down and assisting its evaporation.

The *nails* are horny plates of modified epithelium which protect the tips of the digits. They grow from a root of typical soft epithelial cells at the base of the nail. This root is embedded in the fold of the epidermis. The nails replace the epidermis here, but are continuous with it, so that there is a continuous barrier to exclude bacteria. It is, however, easy for a break in the continuity to occur here if the nails are not carefully looked after, and this is particularly dangerous among nurses, since they are in contact with infection in the course of their work. In animals the nails are protective in more senses than one, but as man has invented implements of various kinds for his use, his nails do not get worn down as do those of the animal, and need regular cutting.

The *sebaceous glands* are small saccular glands which secrete an oily substance called *sebum*. They are situated in the angle between the hair follicle and the arrector pili muscle, so that contraction of the muscle has the effect of squeezing sebum from the gland. This lubricates the skin and hair, and keeps them soft and pliant, so that they do not readily break. However, it picks up dust and bacteria, which cling to an oily surface. As a result we must constantly remove it by washing with soap and water. If some substitute is not applied the skin quickly becomes brittle, so that it breaks readily and lets bacteria enter. The nurse, whose hands are often in lotions and soapy water, which remove the sebum, needs to pay particular attention to this point.

The *functions of the skin* are:

(1) To regulate body temperature.

(2) To excrete waste products.

(3) It is the organ of touch and other senses through which we are made aware of our environment.

(4) It keeps out bacteria by its dry, scaly outer surface.

(5) It secretes sebum.

(6) It protects the body from the harmful effect of the sun's rays by its pigment.

(7) It produces vitamin D through the action of ultra violet rays on the ergosterol it contains.

The regulation of the body temperature. Man is a warm-blooded animal, maintaining a temperature of about $36.9°C$ in spite of variations in the surrounding air. Heat is constantly produced within the body by all the activities of life, and this heat production is balanced by heat loss, so that little rise of temperature occurs.

Heat production is due to the combustion of fuel to produce energy in the tissues. This takes place particularly in the most active tissues, the muscles and the glands. Heat is also absorbed to some extent by the body either through radiation or conduction of heat, as from contact with a hot-water bottle.

Heat loss takes place through many different channels, which in order of their importance are:

(1) Radiation, conduction and convection of heat from the blood in the skin to the surrounding air.

(2) Evaporation of sweat.

(3) Respiration, the expired air being warmed and laden with water vapour.

(4) The excretion of urine and faeces, which is of very slight importance.

The first two are the chief channels of heat loss.

To maintain the normal temperature heat loss must balance heat production, and this is brought about with such accuracy that very little variation occurs in normal health, though there is a slight rise during the day when activity is taking place, and a slight fall during the night when the body is at rest, except for the functioning of vital organs. In normal health the rise during the day is only half a degree C, so that it is so slight as to be negligible.

In *hot weather* to keep the temperature normal:

(1) *Heat production* is lessened, as heat has a depressing effect on the thyroid and adrenal glands, which stimulate tissue activity.

(2) *Heat loss* is increased by:

(a) Dilatation of the blood vessels in the skin, so that loss of heat by radiation, conduction and convection is increased.

(b) Increased sweating, so that more heat is lost by its evaporation.

In *cold weather* to prevent excessive cooling:

(1) *Heat production* is increased, as cold has a stimulating effect on the thyroid and adrenal glands, increasing their secretion.

(2) *Heat loss* is decreased by:

(a) Contraction of the blood vessels in the skin, so that less heat is lost by radiation, conduction and convection.

(b) Decreased sweating, so that less heat is lost in evaporation.

In *exercise*, whatever the weather:

(1) *Heat production* is increased, but this is balanced by:
(2) Increased *heat loss* through:
    (a) Dilatation of the blood vessels in the skin,
    (b) Increased sweating.

In *rest*:

(1) *Heat production* is decreased, but this is balanced by:
(2) Decreased *heat loss* through:
    (a) Contraction of the blood vessels in the skin,
    (b) Decreased sweating.

Exercise is most trying in a hot, wet atmosphere, as the heat prevents loss by radiation, conduction and convection and the moisture already present in the air prevents the evaporation of sweat. This may lead to heatstroke, the body temperature rising excessively even though there has been no exposure to direct sun. Plenty of fluid containing salt should be taken by anyone sweating heavily, or loss of fluid and salts will cause painful muscular cramp.

In man the wearing of clothes and living in warmed houses has lessened the work that the skin is called upon to do, and has resulted in the disappearance of hair, which entangles air, so that it becomes warm and lessens heat loss. The wearing of excessive clothing is not good for health. If the skin is exposed to freely circulating cool air it is able to react to it, and the effect is stimulating and beneficial to health.

The temperature of the body is thought to be controlled by a *heat-regulating centre*, through which heat loss and heat production are balanced. This centre is situated in the hypothalamus, which lies at the base of the cerebrum, just above the pituitary gland, and influences the sympathetic and parasympathetic nervous systems. The centre is probably affected both by the temperature of the blood flowing through it and by the sensory stimuli from the skin and other tissues which are carried to it by sensory nerve fibres. The centre controls both heat production and heat loss. When heat production is increased, or the body is exposed to external heat, it causes dilatation of the blood vessels in the skin through the vasomotor nerves and increased sweating, through the secretory nerves, which control the sweat glands. It also depresses the activity of the adrenal, pituitary, and thyroid glands, through the hypothalamus. When heat production is low, or the body is exposed to external cold, it causes contraction of the blood vessels in the skin and decreased sweating. It also stimulates the adrenal glands, which are closely linked with the hypothalamus, and the pituitary gland, thus having an effect on the endocrine glands generally; this steps up tissue activity and thus heat production. The centre may be affected by disease and injury. Infections, through the action of bacteria and their toxins, generally cause a rise in temperature, called *pyrexia* or fever; this is probably a defence mechanism which makes the tissues less favourable to bacterial growth and reproduction. Tumours, haemorrhage from rupture of a blood vessel at the site, and cranial injury or operation may affect the centre and cause severe hyperpyrexia.

27 The Reproductive Systems

The Male Reproductive System

The male reproductive system consists of:

1. The right and left *testes* and *epididymides*
2. The *deferent ducts*
3. The *seminal vesicles*
4. The *ejaculatory ducts*
5. The *prostate gland*
6. The *penis* and *urethra*

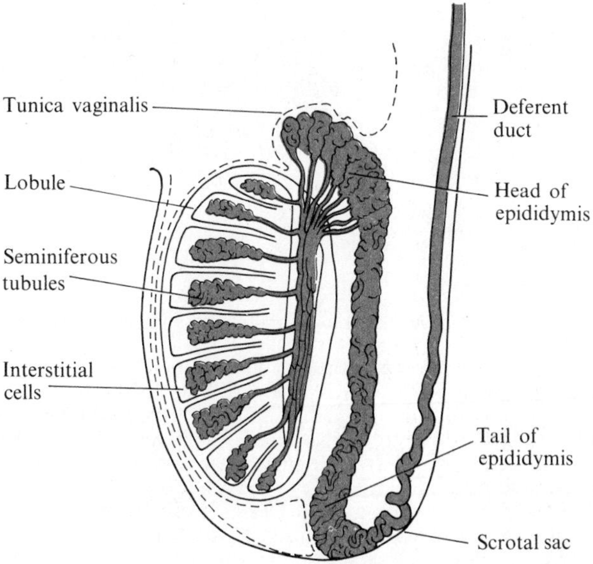

Fig. 205. Structure of the testis.

The *testes* are the glands which produce the male reproductive cells or *spermatozoa*. They are thus the essential male sex organs. They also produce internal secretions.

The *testis* is a gland consisting of 200 to 300 lobules each made up of finely convoluted tubules called the seminiferous tubules; from the lining of these tubules the spermatozoa are produced by a process of cell division. These convoluted tubules lead to the epididymis and carry the spermatozoa there from the testis. There is, in addition, between the tubules gland tissue, called the interstitial cells, which produces the internal secretion of the testes, called testosterone.

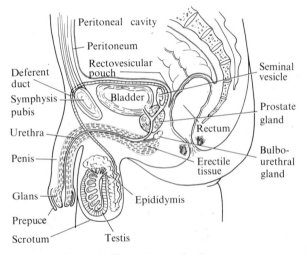

FIG. 206. The male reproductive organs.

The testes lie in the scrotum, a loose pouch of skin hanging from the base of the trunk. They develop in the fetus high up at the back of the abdomen close to the kidneys, and gradually descend through the inguinal canal into the scrotum shortly before birth. It is essential that they should descend into the scrotum, for exposure to the cold external air is necessary for their functioning. The testes normally do not begin to function till the onset of puberty, which is usually at about fourteen to fifteen years. At this age they begin to secrete hormones and produce spermatozoa. Sometimes the testes fail to descend and may remain either in the abdomen or in the inguinal canal. In such cases descent may occur later, but if natural descent into the scrotum does not occur, treatment by gland extracts from the pituitary gland may be attempted to stimulate growth of the ligaments. Operation may be undertaken to bring down the testes and stitch them in position, but as this stretches the blood vessels it may diminish the blood supply and does not always give satisfactory final results; the gland may atrophy if the blood supply is much affected.

The development of the testes at puberty is accompanied by the physical and temperamental developments which characterize the male sex. Hair grows on the face, in the axillae and over the pubis, and the voice breaks as a result of the growth of the larynx and the lengthening of the vocal cords. The characteristic mentality of the male sex appears at the same time. These changes are due to hormones produced by the pituitary gland and testes (see p. 261). In young boys it is quite usual for the testes to be retracted in conditions of cold and anxiety which may prevail in school medical examinations.

As the testis descends it brings down with it a pouch of peritoneum, called the tunica vaginalis. This forms a serous covering for the testis, but the pouch should otherwise be obliterated. If it is not obliterated it forms a possible site for hernia. This causes an inguinal hernia, a loop of bowel or some other organ falling into the pouch which forms the hernia sac.

The *epididymis* is a very fine convoluted tube which is coiled up to form a narrow flattened body lying on the back surface of the testis. The tubules of the testis proper open into it, and it leads in turn to the deferent duct, the duct of the testis. It conveys the reproductive fluid containing the spermatozoa from the testis.

The *deferent duct* runs up from the testis through the inguinal canal and passes over the top of the urinary bladder to the base of the seminal vesicles which lie behind it one on either side. From the testes to the inner end of the inguinal canal the duct is accompanied by the spermatic arteries and veins, and these three structures together form the *spermatic cord*.

The *seminal vesicles* are two small pouches lying at the back of the urinary bladder. They secrete a viscid fluid which forms part of the semen.

The *ejaculatory ducts* lead from the seminal vesicles through the prostate gland to the urethra, which is the common passage for both urine and the semen.

The *prostate gland* lies at the base of the urinary bladder. It is about the size and shape of a chestnut, and the urethra passes through the centre of it. It consists partly of glandular tissue and partly of involuntary muscle. The gland tissue produces a secretion which is thought to act as a lubricant to the spermatozoa. The muscular tissue is the counterpart of the muscle of the uterus of the female, and there is also a minute cavity the counterpart of the uterine cavity.

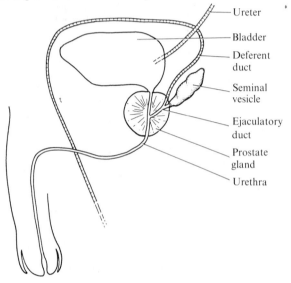

Ureter

Bladder

Deferent duct

Seminal vesicle

Ejaculatory duct

Prostate gland

Urethra

FIG. 207. Relations of seminal vesicles.

The male *urethra*, after running through the prostate gland, passes down the centre of the penis. The penis is composed of a special spongy erectile tissue arranged in three columns through one of which the urethra runs. This tissue is not muscular, but when the veins draining it are constricted it becomes rigid from congestion with blood and is erected. This erection occurs normally during sexual excitement. At the tip of the penis is an enlargement called the glans penis, in the centre of which lies the urinary meatus. The glans is normally covered by a loose double fold of skin called the prepuce or foreskin. It should be possible to draw the foreskin back over the glans penis, but sometimes the opening in it is too small. This is known as phimosis, and is treated either by stretching the fore-skin or by circumcision, i.e. a cutting away of the foreskin, in more severe cases.

The *semen*, or male reproductive fluid, is a composite fluid derived from the testes, the seminal vesicles, and the prostate gland; it contains the spermatozoa.

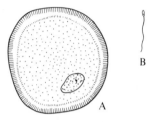

FIG. 208. An ovum and a spermatozöon. (A) Mature ovum; (B) spermatozöon drawn to the same scale.

The *spermatozoa* are minute cells carrying a tail-like projection which joins the cell by a constricted portion called the neck. The tail has a lashing movement, after the semen leaves the male reproductive tract, which enables the cell to move. As a result when the spermatozoa are deposited in the female vagina they can make their way up the uterus and uterine tubes in search of the ova. They are produced in enormous numbers, and it is estimated that on an average 300 000 000 are deposited in the vagina at one time, though only one is necessary to fertilize the ovum.

The Female Reproductive System

The female reproductive system is divided into:

    (1) The external genital organs
    (2) The internal reproductive organs

The *external genital* organs are collectively called the *vulva*, and consist of:

    (1) The mons pubis
    (2) The labia majora
    (3) The perineum
    (4) The labia minora
    (5) The clitoris

(6) The vestibule
(7) The orifice of the urethra
(8) The orifice of the vagina
(9) The hymen
(10) The fossa navicularis
(11) The greater vestibular glands

The *mons pubis* is a pad of fat covered with skin lying over the symphysis pubis. It bears hair after puberty.

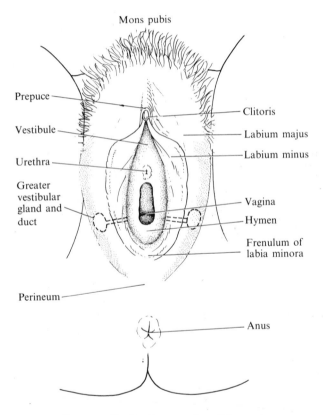

FIG. 209. The female external genital organs.

The *labia majora* are two folds of fatty tissue covered with skin extending backwards from the mons on either side of the vulva and disappearing into the perineum behind. They develop at puberty and are covered with hair on the outer surface after that stage. They atrophy after the menopause.

The *perineum* is the expanse of skin from the vaginal orifice back to the anus. It is about 5 cm in length and bears hair. It lies over the *perineal body*, a mass of muscle and fibrous tissue separating the vagina from the rectum. The muscle of the perineal body is largely the levator ani, the chief muscle of the pelvic floor.

The *labia minora* are two smaller fleshy folds within the labia majora. They meet in front to form a hood-like structure called the *prepuce*, which surrounds and protects the clitoris. The labia minora unite behind in the frenulum of the labia minora (fourchette). This is merely a fold of skin which is often torn in the first labour. The labia minora are covered with modified skin rich in sweat and sebaceous glands to lubricate their surfaces.

The *clitoris* is a small sensitive organ containing erectile tissue corresponding to the male penis. It lies in the front of the vulva immediately below the mons pubis and is protected by the prepuce.

The *vestibule* is the cleft between the labia minora. The orifice of the vagina and the orifice of the urethra open into it.

The *orifice of the urethra* lies at the back of the vestibule, projecting slightly from the normal surface level. At the entrance there are two fine tubular glands, called urethral glands, which secrete lubricating fluid and are important because they tend to harbour infection in cases of gonorrhoea.

The *orifice of the vagina* occupies the space between the labia minora behind the vestibule. It is normally a slit from front to back, the side walls of the vagina being in contact. The orifice is largely blocked in the virgin by the *hymen*. This is a double fold of mucous membrane, usually crescentic in shape, and leaving a gap at the front for the escape of the menstrual flow. Sometimes it has a number of small perforations; this is called a fenestrated hymen. Sometimes there is no opening, but this is usually owing to the vagina not having developed into a canal. The hymen is placed a little inside the orifice of the vagina, so that there is a slight depression between the frenulum of the labia minora and the hymen.

The *greater vestibular glands* are two small glands lying one under each labium majus on either side of the vaginal orifice. Their ducts open laterally to the hymen. They secrete a lubricating fluid to moisten the surfaces of the vulva and so facilitate sexual intercourse.

The whole surface of the vulva is covered with modified skin, i.e. stratified epithelium. It carries hair on the outer surfaces only, but the inner surfaces are exceptionally rich in sebaceous and sweat glands, so that the surfaces are moist and there is no friction on walking.

The *internal reproductive organs* are:

    (1) The ovaries
    (2) The uterine tubes
    (3) The uterus
    (4) The vagina

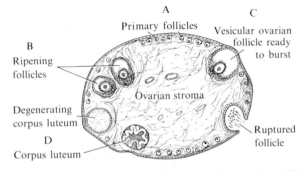

FIG. 210. The ovary, showing: (A) primary follicles; (B) and (C) ripening follicles; and (D) corpus luteum.

The *ovaries* are two small glands about the size and shape of almonds which produce the *ova*. They lie in the pelvic cavity, one on either side of the uterus, to which they are joined by the *ovarian ligaments*. They are attached to the uterine tubes by the *ovarian fimbriae*, and lie on the back of the broad ligaments, which are two double folds of peritoneum extending from the uterus outwards to the pelvic wall at either side. The point of attachment to the broad ligament is called the hilum, and gives entrance to the blood vessels and lymphatics and nerves.

The *cortex*, or outer part of the ovaries, consists of glandular tissue covered by germinal epithelium and contains vast numbers (30 000 to 300 000) of *ova*. Each ovum lies in a little sac of cells called a primary ovarian *follicle*. The ova in their follicles are present at birth, but they are not discharged from the ovary until puberty. Before an ovum is discharged a change takes place in the follicle. The wall develops and ultimately consists of a fibrous outer coat, a vascular coat, a thick coat of many layers of cells from which a projecting mass of cells enclosing the ovum juts inwards. Within the follicle fluid collects which is known as the liquor folliculi. These developments cause a marked enlargement of the follicle, and the fully developed follicle, called the *vesicular ovarian* (or *Graafian*) *follicle*, measures about 1 cm across. These ovarian follicles ripen one at a time, and though at first they pass in from the surface, in the final stages they come to the surface again, project from it, and ultimately rupture through it into the peritoneal cavity, discharging the ovum and the liquor folliculi.

After the rupture of the follicle it fills up with blood clot, which is quickly converted into a mass of specialized tissue called the *corpus luteum*. If the discharged ovum is fertilized this corpus luteum remains till the placenta is formed and takes over its function; later it gradually turns into ordinary scar tissue. If the ovum is not fertilized, the corpus luteum turns into scar tissue in about 6 to 8 weeks.

Both the follicle and the corpus luteum produce *internal secretions*. The hormone of the follicle is called oestrogen, and has a controlling effect on the endometrium, the lining of the uterus, causing a marked thickening or congestion of it. The hormone of the corpus luteum is called *progesterone*, and causes a further thickening and congestion of the endometrium. These changes in the endometrium are essential for pregnancy, the thickened congested endometrium being necessary to supply the growing ovum with the nourishment it needs.

These functions of the ovary are controlled by the *pituitary gland*. The pituitary gland produces two hormones: (a) a follicle-stimulating hormone; and (b) a luteinizing hormone. The follicle-stimulating hormone causes the ripening of the follicle and the production of oestrogen. The luteinizing hormone causes the formation of the corpus luteum and the secretion of progesterone. Normally the follicle stimulating hormone (FSH) and the luteinizing hormone (LH) are secreted in a monthly cycle causing monthly changes in the endometrium, but if the ovum is fertilized and pregnancy follows, a hormone from the ovum causes the pituitary gland to produce the luteinizing hormone so that the corpus luteum persists and maintains the uterus in a condition suitable for childbearing.

The ovaries begin to function at puberty and continue to

discharge ova at monthly intervals from about the age of 13 to about the age of 45 years, the usual time of the menopause. The ovaries are smooth in childhood, but the scars which follow the rupture of the follicles pucker up the surface, so that they finally become very irregular and rather like almonds in appearance as well as shape.

Hormones from the ovaries are also responsible for the development of the reproductive system and the general development which marks puberty in the female, and occurs at about 13 years old. There is marked development of the external genitals, of the uterus and the breasts; hair grows on the genitals and in the axillae; there is a general rounding of the figure and a gradual development of the typical mental outlook of the female personality, which gradually matures as the reproductive organs themselves mature.

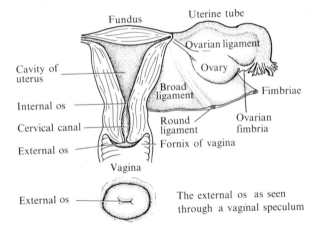

FIG. 211. The female reproductive organs from behind.

The *uterine (Fallopian) tubes* are two fine tubes leading from the peritoneal cavity to the uterus. They are about 10 cm long and are divided into:

(1) The infundibulum, the broad funnel-like end leading from the peritoneal cavity.

(2) The ampulla, thin-walled and tortuous.

(3) The isthmus, a narrow portion joining the ampulla to the uterine part.

(4) A uterine part running through the uterine wall.

The opening into the abdominal cavity is called the *abdominal opening* and is surrounded by fringe-like processes called the fimbriae. One of these fimbriae joins the ovary and is called the ovarian fimbria. Thus the broad mouth of the tube is held lying immediately over the ovary, so that the discharged ova are readily drawn into it.

The *wall* of the uterine tube consists of:

(1) An outer serous covering of *peritoneum*.

(2) A *muscular* coat with peristaltic action.

(3) A lining of *ciliated epithelium*, which is puckered lengthwise to allow a certain degree of stretching of the tube.

The tube conveys the ova from the ovary to the uterus. They are conveyed along the tube partly by the peristaltic action of the muscular wall and partly by the action of the cilia.

It is in the tube that fertilization of the ovum normally takes place. *Fertilization* is the fusion of a spermatozöon and an ovum. The spermatozöon travels up the uterus and along the uterine tube till it strikes and penetrates into an ovum. It then loses its tail and the two cells fuse into one, the fertilized ovum, which develops a tough outer coat to prevent further spermatozoa entering. After fertilization the ovum develops to form the fetus, as the infant is called before birth. One fetus only is usual, but two or more may develop, producing twin or multiple pregnancy. Twins which develop from one ovum are of similar sex and very similar in appearance. Twins may also result from the fertilization of two ova at the same time, but then there is not the same similarity of sex and appearance. If the ovum is not fertilized it passes on into the uterus and is passed out at the next menstrual period. If it is fertilized it begins to multiply and forms a ball of cells which travels along into the uterus and digests a hole, into which it falls, in the thickened endometrium; there it gradually develops to form a sac of membranes with the fetus growing within it, attached by the umbilical cord to the wall of the sac, at a point where the wall has developed to form the *placenta* or afterbirth.

The *uterus* is a pear-shaped muscular organ lying in the centre of the pelvic cavity, with the bladder in front of it and the rectum behind. It measures 7·5 cm in length, 5 cm across, and 2·5 cm in thickness, and weighs about 30 to 40 g.

The upper part is broad and is called the *body* of the uterus. This contains a triangular cavity with the apex downwards. The two uterine tubes lead into the upper angles of the cavity, and the lower angle leads to the cavity of the cervix. The part of the body above the entrance of the uterine tubes is called the fundus.

The lower part is narrow and is called the *cervix*. The cervix contains a canal leading from the cavity of the body into the vagina. The opening from the canal into the uterine cavity is called the *internal os*. The opening from the canal into the vagina is called the *external os*.

The *wall* of the uterus consists of *muscle* covered externally by *peritoneum* and lined with mucous membrane covered with columnar epithelium, called the endometrium. It is 1·25 cm thick.

The muscular wall consists of involuntary muscle fibres arranged in complicated fashion, some longitudinal, some circular, some transverse and some oblique.

The *endometrium* is an epithelial tissue. In the non-pregnant adult it consists of a stroma of irregularly shaped cells in which are set numerous tubular glands. Some of these cells are occasionally ciliated.

The lining of the cervical canal is a typical mucous membrane, rich in large branching mucus-secreting glands which dip deeply into the muscular wall. The lining of the body is a specialized epithelial tissue carrying small tubular glands which secrete fluid, but not typical mucus. These glands only extend through the mucous membrane and do not penetrate the muscular wall. It is from the endometrium lining the body of the uterus that the menstrual discharge occurs each month.

The *blood supply* of the uterus is derived from:

(1) The *ovarian arteries*. These are branches of the abdominal aorta, which run through the broad ligament (see p. 354) and supply the ovary, uterine tube, and fundus of the uterus.

(2) The *uterine arteries*. These are branches of the internal iliac artery which run to the cervix and corkscrew up the wall to anastomose with the ovarian arteries.

The *functions of the uterus* are:

(1) To accommodate the fertilized ovum till the developing child is capable of a separate existence. The fertilized ovum, when it enters the uterus, embeds itself in the lining endometrium, provided this is in a suitable state of congestion and

thickening as a result of the action of the hormones from the follicle and corpus luteum. There it continues to grow till it fills the uterus, and then continues growing, the uterus growing with it during the period of pregnancy. This period lasts for 40 weeks, roughly nine calendar months, and at the end of it the uterus measures 33 × 23 × 20 cm and weighs 1 kg. Meanwhile the single fertilized egg cell has multiplied to form the

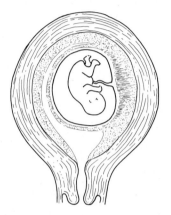

FIG. 212. The uterus at about the tenth week of pregnancy, showing the placenta and developing fetus.

developing fetus, as the infant is called during intra-uterine life, and a bag of membranes filled with fluid in which the infant floats and by which it is protected. The bag of membranes soon fills the uterine cavity and becomes adherent at every point to the uterine wall. At one part of the sac, at the site where the ovum first embedded itself in the endometrium, the placenta develops. The umbilical cord contains blood vessels and runs from the navel of the fetus to the placenta; the placenta receives the mother's blood from the uterine wall and the infant's blood via the umbilical cord. The mother's blood does not pass into the child at any stage, the child developing its own blood and a heart to circulate it as a result of the gradual differentiation of the cells of the developing ovum. In the placenta the child's blood is able to absorb food, oxygen and water from the mother's as it passes through a network of capillaries and gives off its waste products into the mother's circulation. The child is therefore nourished and continues to grow, becoming gradually more and more perfect, and ultimately weighing at the end of the period of pregnancy about 3·2 kg on an average.

(2) To expel the infant when the period of pregnancy is complete. This is termed *labour*, parturition or childbirth, and is brought about by the gradual dilatation of the cervix and vagina and the gradual contraction of the body of the uterus. The dilatation of the cervix enables the child to pass out down the genital tract, and the contraction of the body of the uterus is the driving force which propels the child through it. After the birth of the child the contraction of the uterus separates the membranes and placenta from the uterine wall and expels them also from the birth canal.

(3) *Involution.* This is a gradual return of the uterus and the neighbouring structures to their original size and position. It is due to a process of self-digestion in the uterine tissue. In about eight weeks the uterus is normally almost the same size and shape as before pregnancy.

(4) *Menstruation.* This is the monthly discharge of blood from the endometrium. The endometrium goes through a regular series of changes each month unless pregnancy occurs. Under the influences of oestrogen and progesterone it becomes very much thickened and congested for some days following the rupture of the ovarian follicle. It is then ready to receive the fertilized ovum. If fertilization does not take place the congested blood vessels of the endometrium rupture and blood escapes into the uterine cavity and is expelled by uterine contraction. The endometrium continues to bleed for a period of 4 to 6 days. The normal menstrual blood does not clot and is rich in calcium. It is mixed with fluid from the uterine glands and contains epithelial cells. After the period of menstruation there is a period of repair lasting about 5 days, followed by a period of rest lasting from 10 to 14 days till the premenstrual congestion of the endometrium occurs again. The congestion lasts for about 7 days, and is then again followed by menstruation unless a fertilized ovum has embedded itself in the endometrium.

Position of the uterus. The uterus normally lies in a position of anteversion, with the fundus facing the abdominal wall and the cervix towards the sacrum. There is also slight anteflexion, the body being bent forwards on the cervix at the juncture of these two parts.

Supports of the uterus. The uterus is slung in the pelvis by ligaments and is normally freely movable. The chief ligaments which support it are:

(1) The round ligaments
(2) The lateral, pelvic or transverse ligaments
(3) The uterosacral ligaments

The *round ligaments* run from the fundus of the uterus downwards and forwards through the inguinal canal and are inserted into the deep tissue of the labium majus. They help to keep the fundus tilted forwards, but are comparatively unimportant.

The *transverse ligaments* run from the cervix outwards to the side wall of the pelvis and keep the cervix in the middle line.

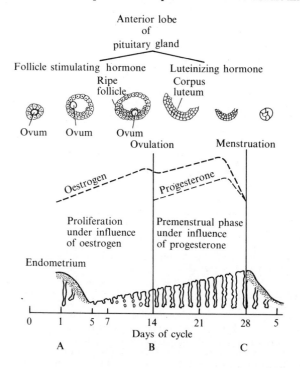

FIG. 213. Diagrammatic representation of the changes in the endrometrium in the menstrual cycle.

They carry the uterine arteries. They are the most important of the uterine ligaments.

The *uterosacral ligaments* run from the cervix backwards on either side of the rectum and attach the cervix to the sacrum.

The *broad ligaments* are two loose double folds of peritoneum running from the uterus outwards to the pelvic wall. They carry to it blood vessels, nerves and lymphatics, but have no supporting value. They cover the uterine tubes which run along the upper edge of the ligament.

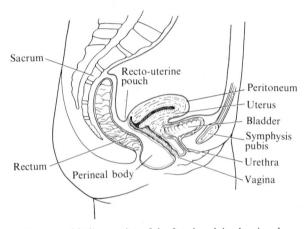

FIG. 214. Median section of the female pelvis, showing the position of the uterus.

The uterus is further supported by the *pelvic floor*, which lies below it, but this does not actually join the uterus at any point, though it does connect with the wall of the vagina which passes through it. This, with the transverse ligaments, forms the main support of the uterus.

The *vagina* is the passage leading from the uterus to the external skin. It runs downwards and forwards from the cervix to the vulva at right angles to the axis of the uterus. The cervix enters its front wall, so that while the back wall is 9 cm long the front measures only 6 to 7·5 cm. The cervix projects into the vault of the vagina, forming pockets all round it, called the *fornices* of the vagina. These are distinguished as anterior, posterior and lateral, according to their position with regard to the cervix.

The wall consists of fibrous and muscular tissue lined with mucous membrane which does not carry any secretory glands. The lining is puckered with both circular and longitudinal puckers which allow for stretching during childbirth. This makes it difficult to cleanse.

The vaginal wall is in relation anteriorly to the bladder and the urethra, and posteriorly to the recto-uterine pouch (the pouch of Douglas), the rectum, and the perineal body. (The recto-uterine pouch is the lowest part of the cavity formed by the reflection of the peritoneum from the back of the uterus over the posterior fornix of the vagina to the rectum.) The vagina allows for the escape of the menstrual flow and of the fetus from the uterus and receives the male organ during coitus.

The Breasts

The *breasts* are two glands which secrete milk and are accessory organs to the female reproductive system. They are present in rudimentary form in the male. They lie on the front aspect of the thorax and vary considerably in size. They are circular in

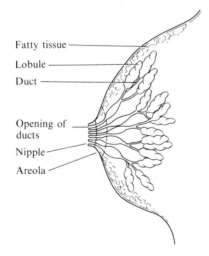

Fatty tissue

Lobule

Duct

Opening of ducts

Nipple

Areola

FIG. 215. The breast.

outline and convex anteriorly. In the centre of the surface is the nipple, which projects normally from the skin level and is pink in the virgin, but pigmented after the first pregnancy.

The breast is a compound saccular gland with its ducts converging to the nipple and opening on its surface in large numbers. The gland tissue is similar to the tissue of the sebaceous gland, but more highly developed and secreting milk in place of sebum. The gland is divided into lobes by partitions of fibrous tissue—a fact which makes the draining of an abscess in the breast difficult. The lobes are subdivided into lobules.

The breasts develop at puberty under the influence of hormones, and further development occurs during pregnancy as a result of hormones from the pituitary gland and ovaries. Fluid known as colostrum is secreted in small amounts by the gland during pregnancy and at the time of child-birth, but the true milk is not secreted till about the third day of the puerperium or lying-in period.

Further Reading

ASIMOV, I. (1959) *The Living River*. Abelard Schuman.

BERRY, R. J. (1965) *Teach Yourself Genetics*. English Universities Press.

CALDER, N. (1970) *The Mind of Man*. B.B.C.

CARTER, C. O. (1962) *Human Heredity*. Pelican.

DAVIS, A. E. and GURR, M. R. (1969) *Blood Groups*. Edinburgh University Press.

FREEMAN, E. H. and BRACEGIRDLE, B. (1966) *An Atlas of Histology*. Heinemann (Educational).

GIBSON, J. (1967) *Guide to the Nervous System*, 2nd edn. Faber.

GREGORY, R. L. (1966) *Eye and Brain*. World University Library.

HARE, R. (1967) *Bacteriology and Immunity for Nurses*, 2nd edn. Longmans.

HARRISON, R. J. (1967) *Reproduction and Man*. Oliver & Boyd.

KILGOUR, O. F. G. (1969) *An Introduction to Physical Aspects of Nursing Science*. Heinemann.

MASON, A. S. (1960) *Health and Hormones*. Pelican.

MUNRO, J. M. (1967) *Pre-Nursing Course in Science*. Livingstone.

RODERICK, G. W. (1968) *Man and Heredity*. Macmillan.

WATSON, J. D. (1968) *Double Helix*. Weidenfeld & Nicholson.

Glossary

New term	Old term
Ampulla, hepatopancreatic	Ampulla of Vater
Apertures, anterior nasal	Anterior nares
Apertures, posterior nasal	Posterior nares
Arch, palatoglossal	Pillars of the fauces
Arch, palatopharyngeal	Pillars of the fauces
Arterial circle of the base of the brain	Circle of Willis
Auricle	Pinna
Bone, spongy	Cancellous bone
Bronchi, principal	Bronchi, main
Capsule, glomerular	Bowman's capsule
Cavity, medullary	Medullary canal
Coeliac trunk	Coeliac axis
Concha, inferior nasal	Turbinate bone
Corium	Dermis
Corpuscle, renal	Malphigian corpuscle
Dens	Odontoid process
Disc, optic	Blind spot
Duct, bile	Common bile duct
deferent	Vasa deferens
nasolacrimal	Nasal duct
Entoderm	Endoderm
Epithelium, pavement	Squamous epithelium
Fenestra cochlea	Round window
Fenestra vestibuli	Oval window
Flexure, left colic	Splenic flexure
Flexure, right colic	Hepatic flexure
Follicle, primary ovarian	Primordial follicle

New term	*Old term*
Follicle, vesicular ovarian	Graafian follicle
Foramen, vertebral	Neural (or spinal) canal
Frenulum of the labia minora	Fourchette
Glands, greater vestibular	Bartholin's glands
suprarenal	Adrenal glands
urethral	Skene's glands
Hypophysis	Pituitary gland
Inferior lobe (of lung)	Lower lobe
Interalveolar cell islets	Islets of Langerhans
Linea terminalis	Pelvic brim
Lymphatic follicle, aggregated	Peyer's patches
Lymphatic trunks	Lymphatic vessels
Macula lutea	Yellow spot
Meatus, external acoustic	External auditory canal
Membrane, tympanic	Eardrum
Mons pubis	Mons veneris
Muscle, sternocleidomastoid	Sternomastoid muscle
striped	Striated (voluntary) muscle
unstriped	Plain (involuntary) muscle
Nerve, accessory	Spinal accessory nerve
Nerve, vestibulocochlear	Auditory nerve
Palatine bones	Palatal bones
Pelvis, greater	False pelvis
Pelvis, lesser	True pelvis
Pouch, recto-uterine	Pouch of Douglas
Process, xiphoid	Xiphisternum (or ensiform cartilage)
Pulmonary trunk	Pulmonary artery
Sinus, lateral	Transverse sinus
Skull cap	Vertex (or vault) of skull
Sulcus, central	Fissure of Rolando
Sulcus, lateral	Fissure of Sylvius
Superior lobe (of lung)	Upper lobe

New term	*Old term*
Tonsil, palatine	Tonsil
Tonsil, pharyngeal	Adenoid
Tube, auditory	Eustachian tube
renal	Uriniferous tubule
uterine	Fallopian
Veins, brachiocephalic	Veins, innominate
Zygomatic bone	Malar (or cheek) bone

Index

Baillière's Nursing Books

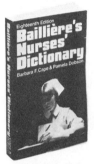

Baillière's Nurses' Dictionary

Cape & Dobson

The 18th edition of this famous dictionary has been completely reset and updated in a new format. It contains hundreds of new definitions, radically revised appendices and new illustrations — an informative, pocket-sized dictionary for student and trained nurse alike.

1974 • 18th edn. • Limp.

Baillière's Midwives' Dictionary

Da Cruz & Adams

The ideal pocket-sized dictionary for midwives and obstetric nurses. "A little mine of invaluable information…it really does contain the exact definition wanted in a hurry."
Midwives' Chronicle 1976 • 6th edn. • Limp.

Baillière's Pocket Book of Ward Information

Fully revised and brought up-to-date, this book contains a multitude of useful information likely to be needed by nurses in their day to day work and of particular help to nurses in training.

1971 • 12th edn. • Limp.

The Nurses' Aids Series

The Nurses' Aids Series is planned to meet the needs of the student nurse during training, and later in qualifying for another part of the Register, by providing a set of textbooks covering most of the subjects included in the general part of the Register and certain specialist subjects. The pupil nurse, too, will find many of these books of particular value and help in practical bedside training. The Series conforms to three factors important to the student.

1. All the authors are nurses who know exactly what the student requires.
2. The books are frequently revised to ensure that advances in knowledge reach the student as soon as practicable.
3. The Aids are well printed and easy to read, clearly illustrated, and modestly priced.

Anaesthesia & Recovery Room Techniques/Wachstein
1976 • 2nd edn.

Anatomy & Physiology for Nurses/Armstrong & Jackson
1972 • 8th edn.

Arithmetic in Nursing/Fream & Davies
1972 • 4th edn.

Ear, Nose & Throat Nursing/Marshall & Oxlade
1972 • 5th edn.

Geriatric Nursing/Storrs
1976 • 1st edn.

Medical Nursing/Chapman
1977 • 9th edn.

Microbiology for Nurses/Bocock & Parker
1972 • 4th edn.

Obstetric & Gynaecological Nursing/Bailey *1976 • 2nd edn.*

Ophthalmic Nursing/Darling & Thorpe *1975 • 1st edn.*

Orthopaedics for Nurses/ Davies & Stone *1971 • 4th edn.*

Paediatric Nursing/Duncombe & Weller *1974 • 4th edn.*

Personal & Community Health/ Jackson & Lane *1975 • 1st edn.*

Pharmacology for Nurses/ Bailey *1975 • 4th edn.*

Practical Nursing/Clarke *1977 • 12th edn.*

Practical Procedures for Nurses/Billing *1976 • 2nd edn.*

Psychiatric Nursing/Altschul & Simpson *1977 • 5th edn.*

Psychology for Nurses/ Altschul *1975 • 4th edn.*

Sociology for Nurses/Chapman *In preparation.*

Surgical Nursing/Fish *1974 • 9th edn.*

Theatre Technique/Houghton & Hudd *1967 • 4th edn.*

Tropical Hygiene & Nursing/ Fream *New edition in preparation.*

NAS Special Interest Texts *See over*

For the Advanced Student

NURSES' AIDS SERIES SPECIAL INTEREST TEXTS

Special Interest Texts will enable the student nurse to study a particular subject in greater detail during training or after basic studies have been completed.

Gastroenterological Nursing/ Gribble   *1977 • 1st edn.*

Neuromedical & Neurosurgical Nursing/Purchese   *1977 • 1st edn.*

Baillière's Medical Transparencies

* A visual reference library for lecturers and students.

*Of special interest to Nurses and Nurse Tutors are 'BMT 1', on the Anatomy of the Head, Neck and Limbs and 'BMT 2', on the Anatomy of the Thorax and Abdomen. Each set illustrates the major anatomical features of the regions with 21 and 18 slides in full colour. Other sets of interest to nurses specializing in these topics are **Paediatrics** 'BMT 17' and **Venereal Diseases** 'BMT 9' together with **Other Sexually Transmitted Diseases** 'BMT 19'. 24 slides in each set.

Quizzes and Questions for Nurses

Book A. Medical Nursing and Paediatric Nursing

Book B. Surgical Nursing and Geriatric Nursing

E. J. Hull and B. J. Isaacs

Two new books devised specially to help nurses in training to revise for examination purposes the subjects that they have been studying. The books are similar to the successful 'Do-It-Yourself' Revision series, but are set at a more elementary level. Each chapter is based on an examination question. The 'revision' part of each chapter consists of a quiz, with answers and explanations, covering the material to be revised.

1976 • Book A • 160pp. • 16 Illus. • Limp
Book B • 160pp. • 10 Illus. • Limp

Do-It-Yourself Revision For Nurses Books 1, 2, 3, 4, 5 & 6

E. J. Hull and B. J. Isaacs

The six books of this series provide a comprehensive framework for revision of the GNC syllabus and developments made since to it. The student reviews a subject of choice, answers questions selected from recent State Final Examinations, and marks her replies against the model answers provided.
'Highly recommended to all student nurses as a planned guide to revision.' *Nursing Times.*

*1970-1972 • **Books 1-6** • 135pp average •*
illustrated

Standard Textbooks

Ward Administration & Teaching/Perry

"This is a book which has long been needed. Every trained nurse could learn something from it." *Nursing Mirror*   1968 • 1st ed

Handbook of Practical Nursing/Crispin

This replaces the well-known textbook by Swire and is for the pupil and student nurse, written in a clear, easy-to-read style.   1976 • 1st ed

Nursery Nursing — A A Handbook for Nursery Nurses/Meering& Stacey

"…a valuable reference book for all nurses interested in nursery work…" *The Lamp*
1971 • 5th ed

Basic Nursing Care A Guide for Nursing Auxiliaries/Hutton

A practical, induction period, reference book for nursing auxiliaries in association with in-service training schemes and ward procedure manuals.
1974 • 1st ed

Nursing in the Community/ Keywood

Provides a comprehensive account of the various problems peculiar to nursing outside the hospital, and describes the organization of a modern community nursing service.

1977 • 1st ed

Books for the Psychiatric Nurse

Psychiatric Nursing/Altschul
Indispensable to students training for admission to the Register of Mental Nurses.
1977 • 5th edn. • 400pp. • Limp

Nursing the Psychiatric Patient/ Burr & Budge
For students in psychiatric training.
1976 • 3rd edn. • 320pp. • 15 illus. • Limp

Clinical Aspects of Dementia/ Pearce & Miller
Focuses on the treatable presenile dementias.
"Extremely useful" — Nursing Mirror
1973 • 160pp. • 12 plates, 15 illus.

Books for the Midwife

Mayes' Midwifery/Bailey
1976 • 9th edn. • 530pp. • 180 illus. • Limp

Obstetric & Gynaecological Nursing/Bailey
(Nurses' Aids Series)
1975 • 2nd edn. • 343pp. • 131 illus. • Limp

For Nurses Working in Intensive Care units

Patient Care: Cardiovascular Disorders/Ashworth & Rose
1973 • 309pp. • 110 illus.

Nurses' Guide to Cardiac Monitoring/Hubner
1975 • 2nd edn. • 66pp. • 37 illus. • Limp.

Cardiology/Julian
1973 • 2nd edn. • 341pp. • 112 illus. • Limp

Nursing Care of the Unconscious Patient/Mountjo & Wythe
1970 • 104pp. • 11 illus. • Limp

A complete catalogue and current price lists are available on request direct from the publishers

BAILLIÈRE **TINDALL**

35 Red Lion Square, London WC1R 4SG

The details and editions in this list are those current at the time of going to press but are liable to alteration without notice.